Drawing Standards for Computer-aided Engineering

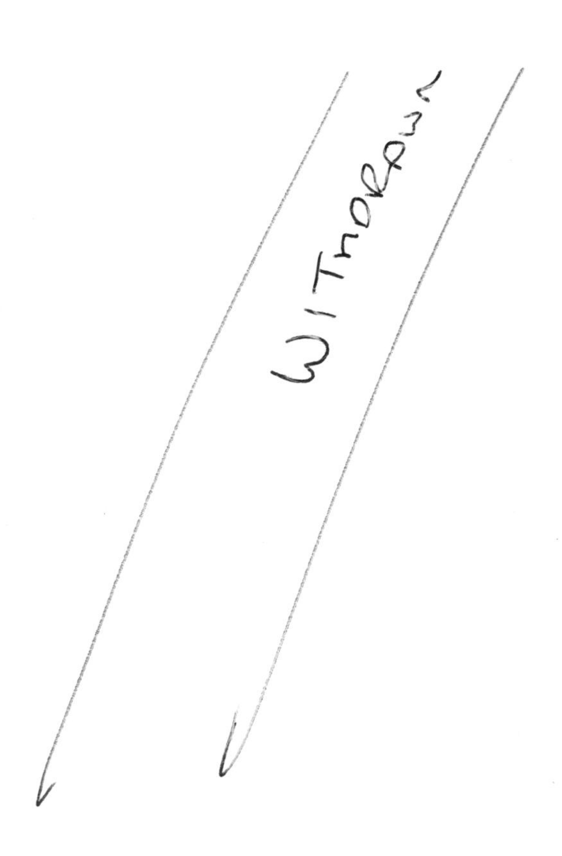

City and Guilds Co-publishing Series

City and Guilds of London Institute has a long history of providing assessments and certification to those who have undertaken education and training in a wide variety of technical subjects or occupational areas. Its business is essentially to provide an assurance that pre-determined standards have been met. That activity has grown in importance over the past few years as government and national bodies strive to create the right conditions for the steady growth of a skilled and flexible workforce.

Both teachers and learners need materials to support them as they work towards the attainment of qualifications, and City and Guilds is pleased to be working with several distinguished publishers towards meeting that need. It has been closely involved in planning, author selection and text appraisal, although the opinions expressed in the publications are those of the individual authors and are not necessarily those of the Institute.

City and Guilds is fully committed to the projects listed below and is pleased to commend them to teaching staff, students and their advisers.

Carolyn Andrew and others, *Business Administration Level I* and *Business Administration Level II*, John Murray
Chris Cook, *Assessor Workbook*, Macmillan
David Minton, *Teaching Skills in Further and Adult Education*, Macmillan
Graham Morris and Lesley Reveler, *Retail Certificate Workbook* (Levels 1 and 2), Macmillan
Peter Riley (consultant editor), *Computer-aided Engineering*, and associated Workbooks: *CNC Setting and Operation*; *CNC Part Programming*; *Computer-aided Draughting*; *Robot Technology*; *Programmable Logic Control*; *Drawing Standards for Computer-aided Engineering*, Macmillan
Barbara Wilson, *Information Technology: the Basics*, Macmillan
Caroline Wilkinson, *Information Technology in the Office*, Macmillan

Drawing Standards for Computer-aided Engineering

M. A. Parker, MEd, IEng, MIMechIE

Formerly Senior Technical Education Officer, British Standards Institution

First published 1995 by
MACMILLAN PRESS LTD
Houndmills, Basingstoke, Hampshire RG21 6XS
and London
Companies and representatives
throughout the world

ISBN 0–333–62532–3

10 9 8 7 6 5 4 3 2 1
04 03 02 01 00 99 98 97 96 95

A catalogue record for this book is available
from the British Library

Typeset by Tek Art, Croydon, Surrey

Printed and bound in Great Britain by
Biddles Ltd,
Guildford and Kings Lynn

Contents

Introduction

The use of computer-aided draughting systems in all areas of the engineering industry is rapidly increasing as the cost of the systems falls and their sophistication increases. They can produce drawings of all types much faster and more simply than is possible using the traditional pencil, instruments and drawing board, they can be integrated with the manufacturing process, and they can provide savings of time and money which help to reduce the cost of the finished product.

Nevertheless, however drawings are produced, they must conform to the conventions and methods of presentation laid down in national and international standards. If they do not, then confusion can arise and time and money can be wasted, particularly if the drawings are exchanged internationally. The aim of this workbook is to present and explain those methods and conventions which are used most often.

The British Standard which deals with engineering drawing practice is BS 308 and computer-aided draughting machines generally are capable of complying with it. Any minor exceptions due to limitations of a particular system should not lead to misunderstanding of a drawing which is claimed to comply with that standard.

The British Standards Institution has kindly given permission for extracts from BS 308 : Part 1 *Recommendations for General Principles* and Part 2 *Recommendations for the Dimensioning and Tolerancing of Size* to be included, together with material from BS 4500 *ISO Limits and Fits*. Copies of these and other standards mentioned in the text can be obtained from BSI, Publications Department, 389 Chiswick High Road, London W4 4AA.

City and Guilds/Macmillan publishing for computer-aided engineering

This workbook is complementary to a series of City and Guilds/Macmillan books which together give complete and up-to-date coverage of computer-aided engineering. A core text, or source book (*Computer-aided Engineering*), gives basic information on all the main topic areas (basic CNC; CNC setting and operation; CNC part programming; CNC advanced part programming; basic CAD/CAM; computer-aided draughting; advanced CAD; basic robotics; robot technology; programmable logic controllers; more advanced programmable logic controllers). It has tasks structured into the text to encourage active learning.

Workbooks cover five main topics: CNC setting and operation; CNC part programming; computer-aided draughting; robot technology; programmable logic controllers. Each workbook includes all the operational information and guidance needed for completion of the practical assignments and tasks.

The books complement each other but can be used independently of each other.

Peter Riley (formerly Head of Department of Engineering Technology, Blackpool and The Fylde College) is Consultant Editor of the series.

How to use this book

Each learning assignment in this workbook has a similar structure to make its use as straightforward as possible. Information and guidance that is needed for the completion of the practical work is included with each assignment.

You will be able to identify the following parts of the text:

- Background information introducing the topic at the beginning of each assignment.
- Other relevant knowledge given under the heading 'Additional information'.
- The practical 'Tasks' are presented in a logical sequence. In many cases the later tasks in an assignment are included to reinforce and enhance the basic practical work.
- If there is information of particular interest concerning the practical tasks, you will find this under the heading 'A point to note' or 'Points to note'.

All the diagrams which are needed for each assignment are given at the appropriate point in the text.

You are recommended to obtain a folder in which to keep work which you have completed. This will serve as a record of your achievements and may be useful for future reference.

Learning Assignment 1

Orthographic projection

Engineering drawing is the language of engineers. They use it to provide information about engineering parts to everyone who is concerned in their design, manufacture and inspection. All the information which is needed to make the parts must be given clearly and precisely and in a way which is universally understood. This can be achieved by producing drawings which follow the recommendations of British Standard (BS) 308 *Engineering Drawing Practice.*

All drawings of three-dimensional objects, whether artistic or technical, must solve satisfactorily the fundamental problem of representing them on a two-dimensional sheet of paper. This problem is generally solved in engineering by using drawings in orthographic projection. Such drawings are relatively easy to prepare and give accurate information about the size and shape of a part. However, to read and interpret them fluently does require some practice and experience.

A POINT TO NOTE

When a line is parallel to the plane on which it is projected orthogonally, the projection will show the true length of the line; see Figure 1.2(b). Similarly, when an area is projected onto a plane to which it is parallel, the projection will be the true shape of the area; see Figure 1.3(c). In Figure 1.4(a) and (b) a face of the solid is parallel to the projection plane, so that face will appear as its true shape and its boundaries will be true lengths.

Orthogonal projectors

Figure 1.1 shows three points A, B and C, each being next to a plane. The lines from the points to the planes, meeting them at a, b and c, are projectors and a, b and c are the projections of A, B and C on the planes. If the projectors are perpendicular (normal) to the planes then they are orthogonal, which means 'at right angles', and a, b and c are orthogonal projections of A, B and C.

A line can be projected orthogonally by projecting its ends, as shown in Figure 1.2, the projectors being parallel and at right angles to the plane. In

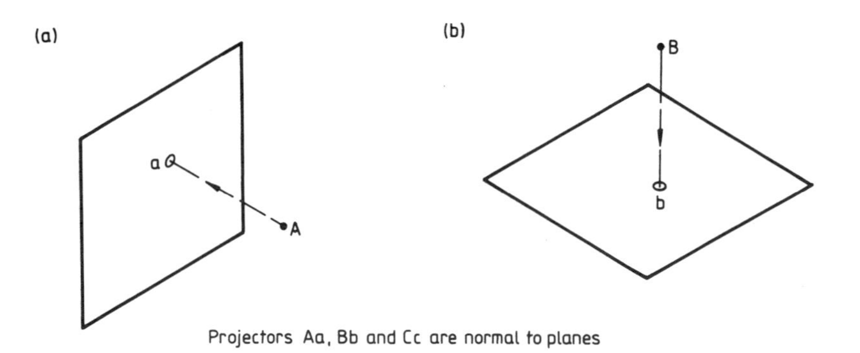

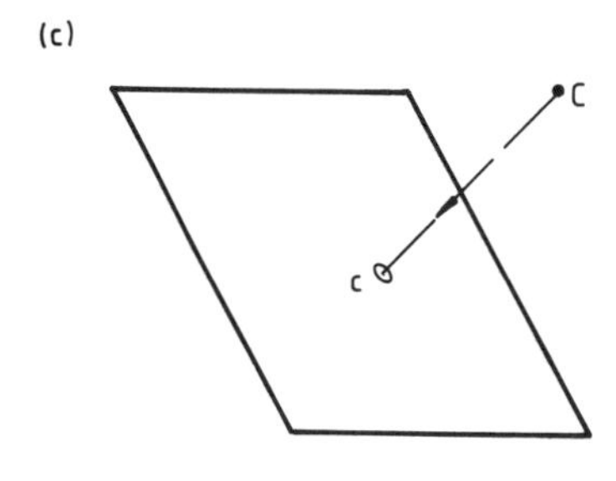

▲**Figure 1.1** Orthogonal projection of a point

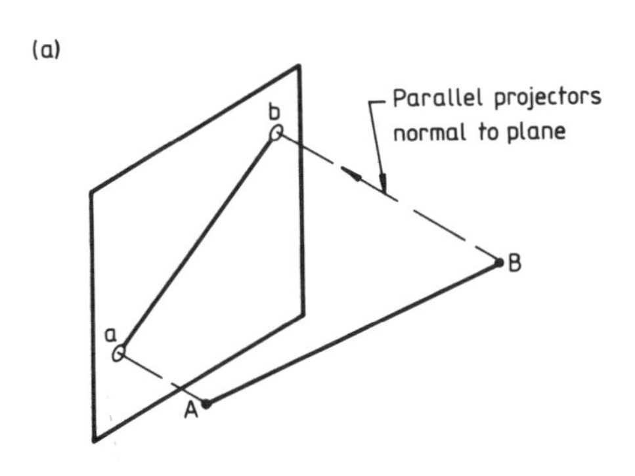

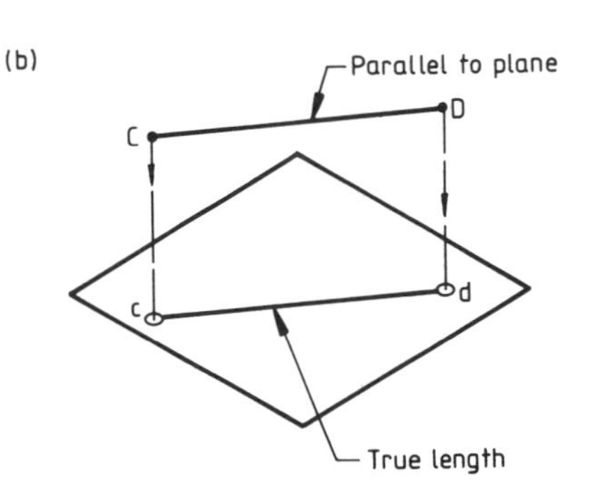

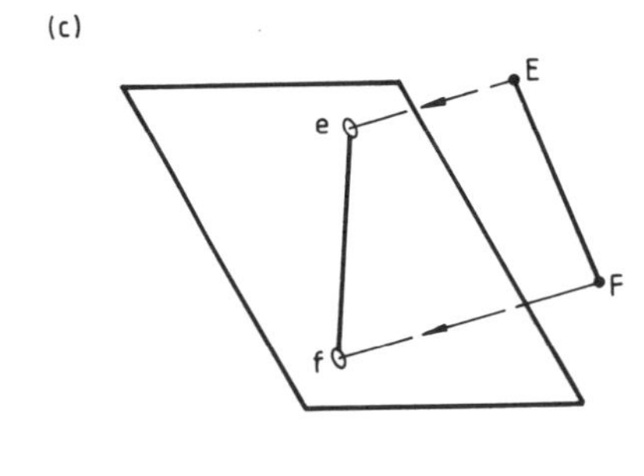

▲**Figure 1.2** Orthogonal projection of a line

Figure 1.3 the lines bounding an area are projected orthogonally, so producing an orthogonal projection of the area.

The faces of a solid are areas, so by using sets of parallel projectors the solid can be projected onto any number of planes. This is illustrated in Figure 1.4. As before, the projectors in each set are parallel and at right angles to their projection plane.

▼**Figure 1.3** Orthogonal projection of an area

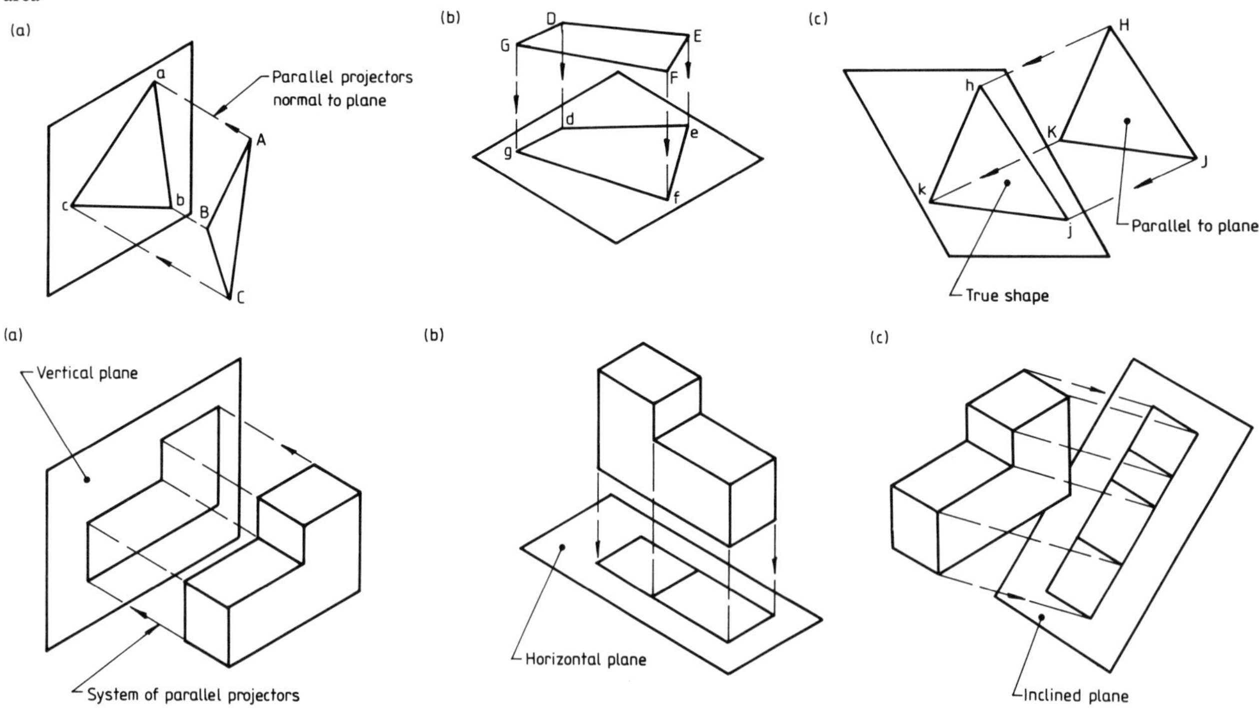

▲**Figure 1.4** Orthogonal projection of a solid

Orthographic projection

Views of an object projected orthogonally onto a vertical plane (VP) and a horizontal plane (HP) are, when taken together, said to be in orthographic projection. The vertical and horizontal planes are the principal planes of projection. They intersect producing four quadrants or angles, as shown in Figure 1.5, but only the first and third angles are used in practice to produce drawings.

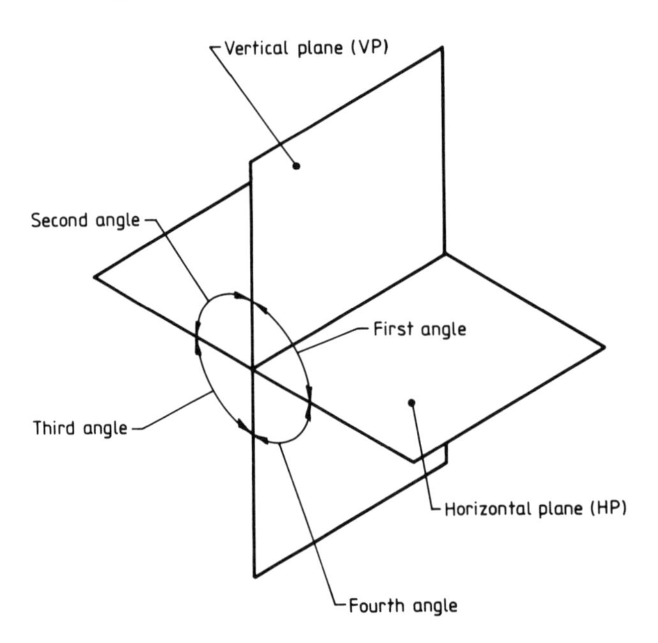

►**Figure 1.5** Principal planes of projection

The object to be drawn is imagined to be placed in one of these quadrants, and orthogonal views of it are projected onto the planes. These are the principal views of the object. The object may be placed in any position relative to the planes, but usually is arranged with its main faces parallel to them. As explained earlier, the faces then project as their true shapes and the boundaries of the projections are true lengths. Therefore dimensions applied to the views will be true lengths and the views will give an accurate picture of the size and shape of the object.

First-angle orthographic projection

In Figure 1.6(a) an object is shown positioned in the first quadrant. Orthogonal views of it have been drawn on the planes. The view on the vertical plane is the elevation, that on the horizontal plane is the plan. To show the views in the positions they would occupy on a sheet of paper, as in Figure 1.6(b), the horizontal plane is opened out in the direction of arrow S about the intersection line between the planes. This line is called the XY line, ground line or folding line. Relative to the elevation it represents the horizontal plane; relative to the plan it represents the vertical plane.

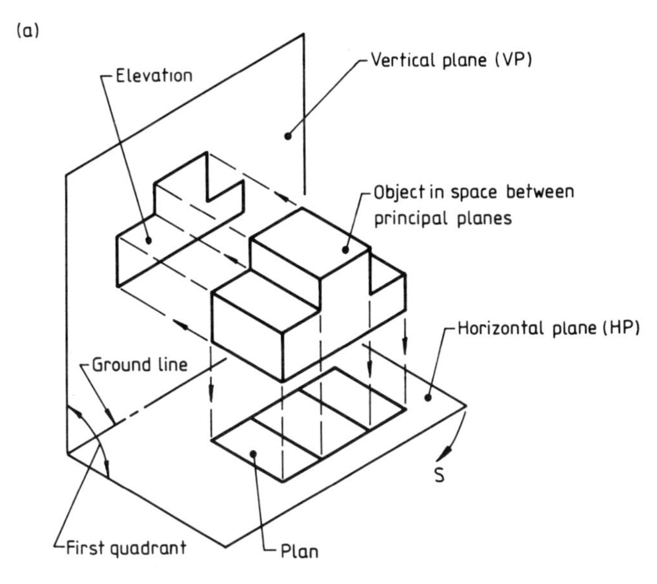

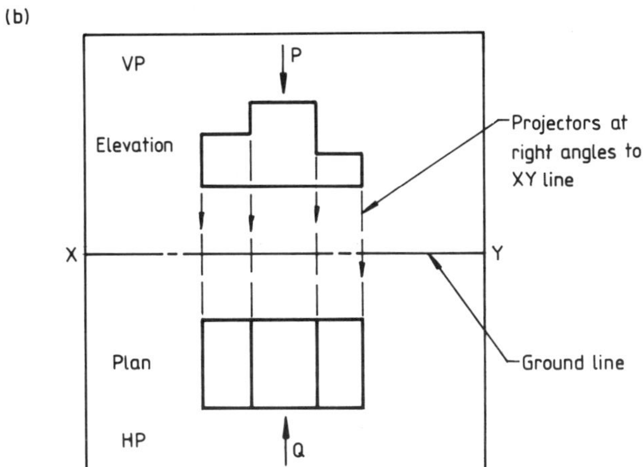

◀**Figure 1.6** Principal views in first-angle projection

From Figure 1.6 it can be seen that:

(a) In first-angle projection the elevation appears above the plan.
(b) The elevation is as far above the XY line as the object is above the horizontal plane.
(c) The elevation is obtained by viewing the plan in the direction of arrow Q.
(d) The plan is as far below the XY line as the object is from the vertical plane.
(e) The plan is obtained by viewing the elevation in the direction of arrow P.

> *A POINT TO NOTE*
> A drawing prepared for use in industry will not show the vertical and horizontal planes, the XY line or the projectors. They are shown here to explain the underlying principles of the projection method.

Third-angle orthographic projection

An object positioned in the third quadrant is shown in Figure 1.7. The planes now come between the observer and the object, so they are

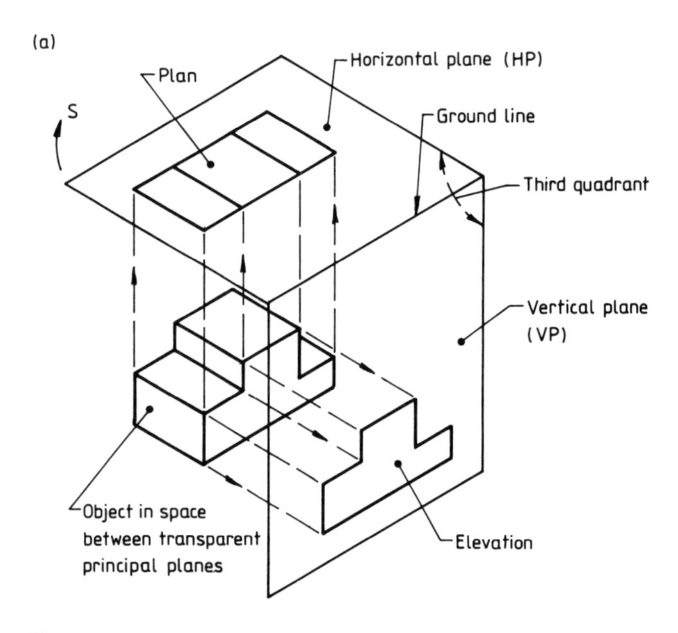

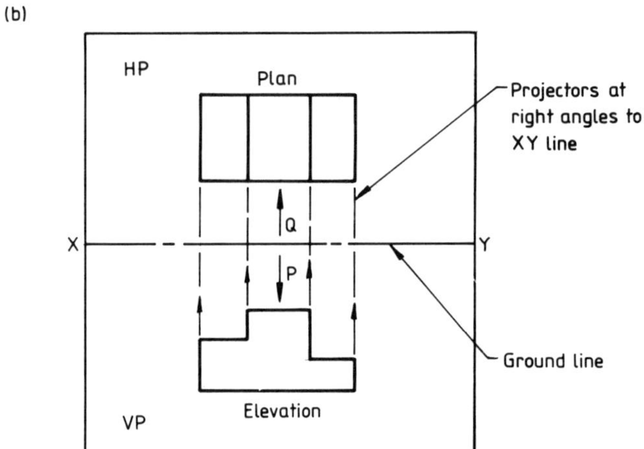

▶**Figure 1.7** Principal views in third-angle projection

considered to be transparent. An elevation and plan have been projected orthogonally onto the planes, and the horizontal plane has again been opened out about the XY line in the direction of arrow S.

The views now appear as in Figure 1.7(b), which shows that:

(a) In third-angle projection the plan is drawn above the elevation.
(b) The plan is as far above the XY line as the object is from the vertical plane.
(c) The plan is obtained by viewing the elevation in the direction of arrow P.
(d) The elevation is as far below the XY line as the object is below the horizontal plane.
(e) The elevation is obtained by viewing the plan in the direction of arrow Q.

A POINT TO NOTE

The views of an object are the same whether they are drawn in first- or third-angle projection. Only their relative positions are different.

Projection symbols

Both first-angle and third-angle projections have been standardized and are recognised internationally. To avoid possible costly confusion the system which has been used must be indicated on the drawing. This is done by a symbol consisting of two views of a frustum of a cone. Figure 1.8(a) shows

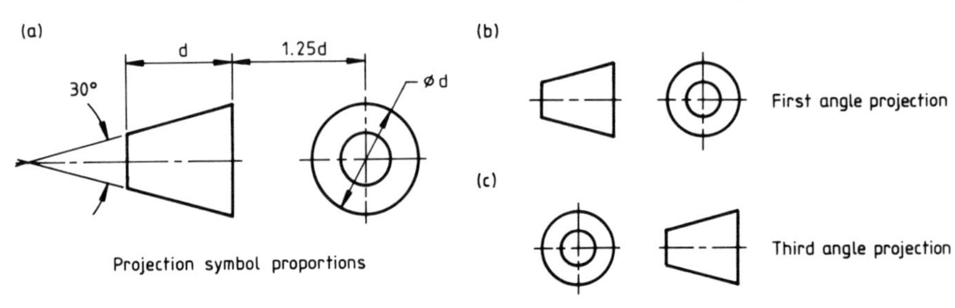

▶**Figure 1.8** Projection symbols

the proportions of the symbol which are recommended by BS 308 and Figure 1.8(b) and (c) the symbols for first- and third-angle projections, respectively.

A POINT TO NOTE

All drawings in orthographic projection produced for the tasks in this workbook may be drawn in either first- or third-angle projection, unless the required projection angle is specified. In all cases the correct projection symbol must appear on the drawing.

The projection of end views

Although many engineering parts can be described completely with two views, and some indeed with only one, there are cases where a third view is needed to show all the details of the part or to avoid ambiguity. The third view is called an end view, or sometimes an end elevation or a side elevation.

The end view is drawn on an auxiliary vertical plane which is at right angles to both the horizontal and vertical planes. The plane may be placed in either of the positions shown in Figure 1.9(a), which shows the first quadrant. The position chosen in any particular case depends on which of the end faces has detail which must be shown for the complete description of the part. Sometimes it is necessary to show both end views.

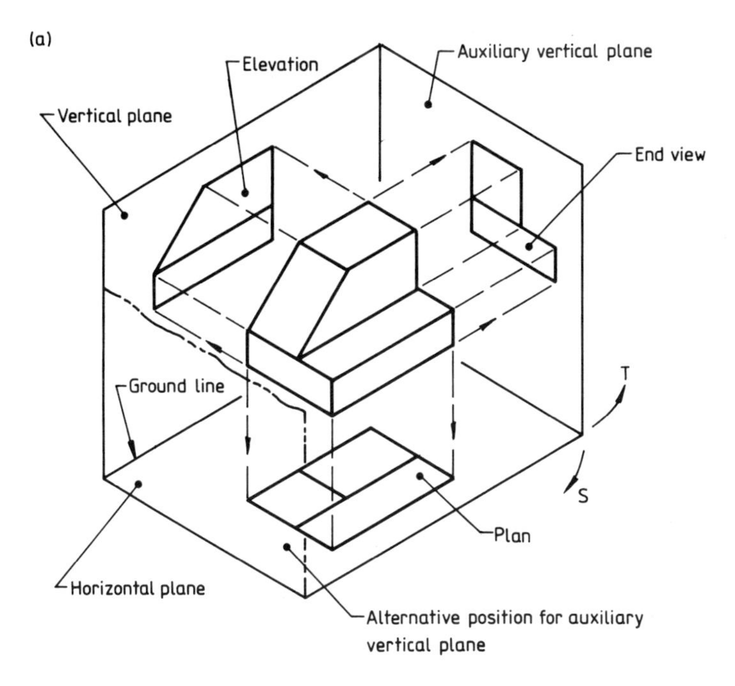

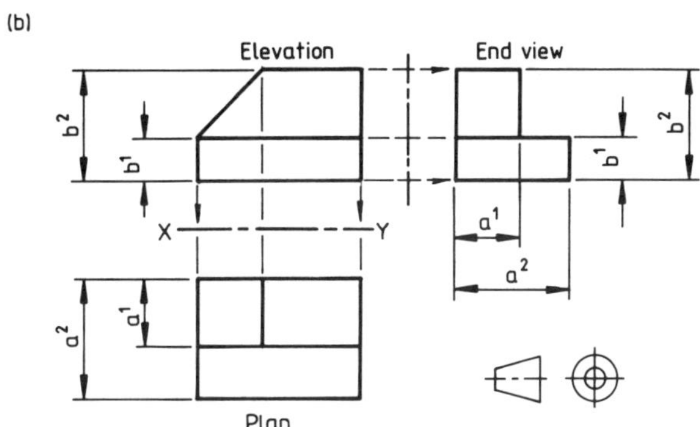

◀**Figure 1.9** Projection of end views in first-angle projection

The horizontal plane is opened out in the direction of arrow S, as before, and the auxiliary vertical plane in the direction of arrow T. The three views then occupy the positions shown in Figure 1.9(b). It can be seen that:

(a) Heights b^1 and b^2 in the elevation and the end view are equal.

(b) Depths a^1 and a^2 in the plan are equal to the corresponding widths in the end view.

The two possible positions of the auxiliary vertical plane in the third quadrant are shown in Figure 1.10(a). Figure 1.10(b) shows the positions of the three views after the horizontal and auxiliary vertical planes have been opened out. This figure again shows that corresponding heights in the elevation and end view are equal, and that depths in the plan view are equal to corresponding widths in the end view.

POINTS TO NOTE

- When orthographic projection is used to make a drawing of an object, each view shows only two of the three dimensions – length, breadth and height – of the object. The third dimension appears in an adjacent view.
- If any two of the three views – elevation, plan and end view – of an object are available, the third can be constructed from them.

POINTS TO NOTE

- Before starting a drawing it is essential to have a clear mental picture of the views required and their positions on the paper. If the part is complex then a sketch of possible views will often clarify which ones should be used and where they should be placed.
- Consider whether the drawing should be made to a scale other than full size. Scales are dealt with in Assignment 2.
- Always select a standard size drawing sheet which will provide enough space for the views and accommodate dimensions and notes between and around them without crowding. If there is any doubt about the space required, use the next larger size of drawing sheet.
- Use the minimum number of views consistent with describing the part completely. A view which shows only a diameter or a thickness is unnecessary if this information can be given as a note or a dimension on another view. A view without a note or dimensions is probably unnecessary.
- Build up all the views together; completing them separately wastes time. Measurements can often be made on two or more views simultaneously, or projected or transferred with dividers from one to another immediately.

Comparing Figures 1.9(b) and 1.10(b) shows that in first-angle projection the plan and end views are placed at the sides of the elevation furthest from the faces they represent. In third-angle projection they are placed nearest to the faces they represent. So for long objects third-angle projection has the advantage that end views appear nearest to the faces they describe.

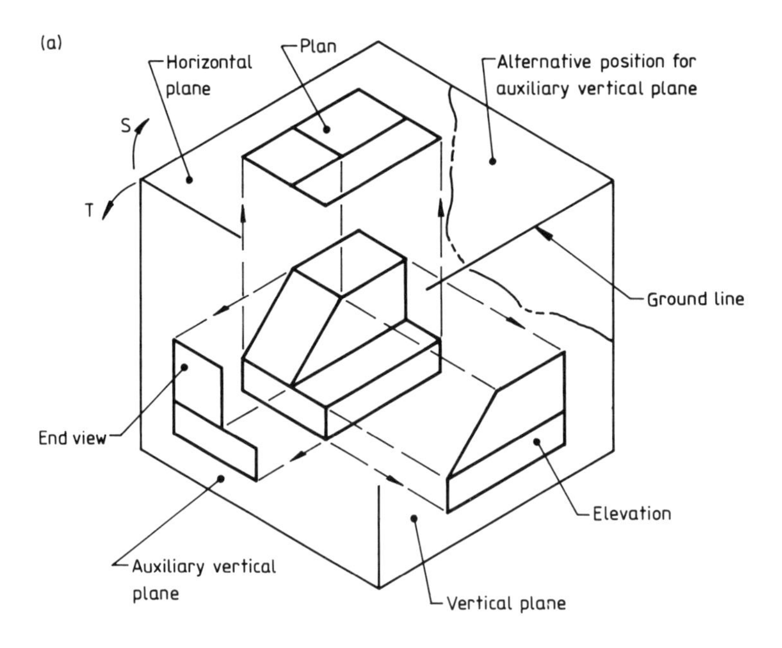

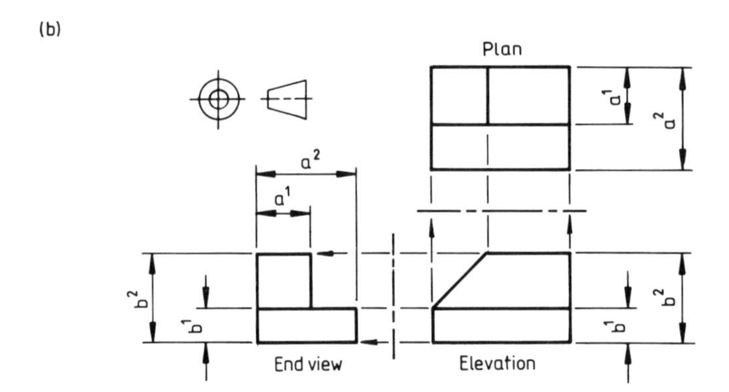

▶Figure 1.10 Projection of end views in third-angle projection

Additional information

Further orthographic views of an object can be drawn by projecting it onto any number of auxiliary planes. The projectors must be parallel to each other and normal to the plane of projection.

Special arrangement of views

In exceptional circumstances it may be impossible to draw a view in the position required by the projection symbol. In this case the view may be drawn anywhere on the drawing which is convenient, but the direction of viewing must be shown by an arrow and the resulting view must be identified.

Outlines of views

Different types of lines are used on drawings and that for the outlines and edges of views is a continuous thick line. Because almost all other lines are thin, this ensures that the most prominent lines on a drawing are the outlines, thus emphasizing the shape of the part. Thick lines are twice as wide as thin lines, and commonly used widths are 0.7 mm and 0.35 mm.

Additional information

The direction in which lines are drawn using a CAD system may be significant if the part is to be made using a CAM or CNC machining program because the direction of the cutter path often follows the direction of the drawn profile.

Intersection points on a screen at different elements of a profile need to connect mathematically to ensure continuity of the tool path in any subsequent machining program.

TASKS

Task 1.1

Draw the following undimensioned views of the milled blocks in Figure 1.11:

- An elevation in the direction of arrow A.
- A plan in the direction of arrow B.
- An end view in the direction of arrow C.

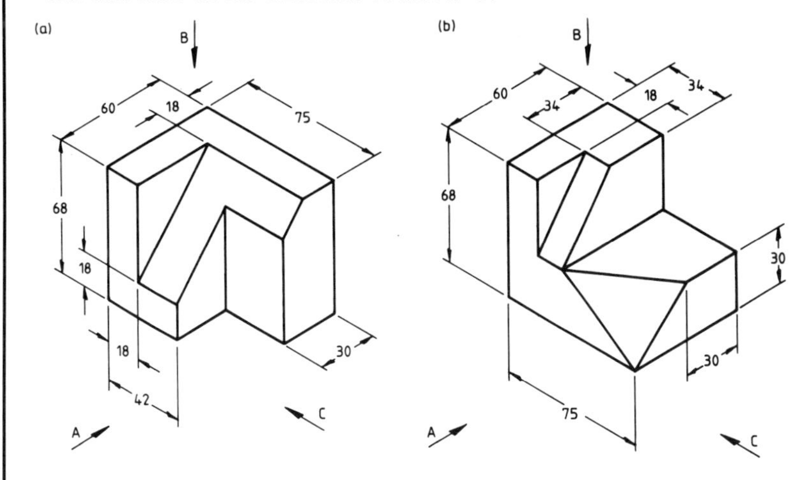

◀**Figure 1.11** Milled blocks

Use first-angle projection for the block at (a) and third-angle projection for the block at (b).

Task 1.2

Draw the third orthographic view where shown for the four objects in Figure 1.12. Dimensions are not required. Drawings (a) and (b) are in first-angle projection and (c) and (d) in third.

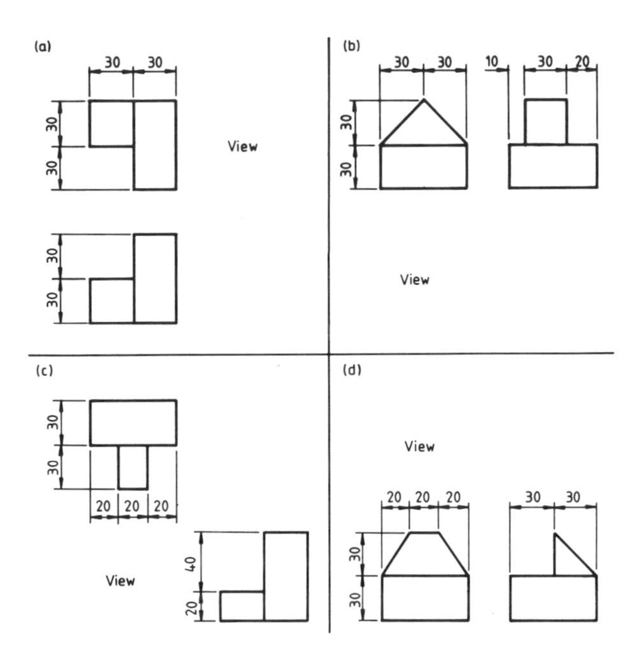

◀**Figure 1.12** Projection of a third orthographic view

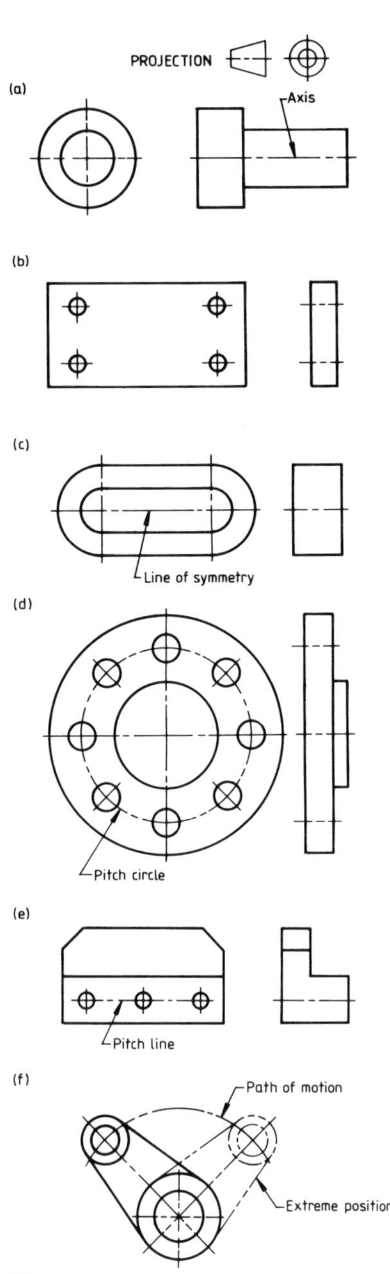

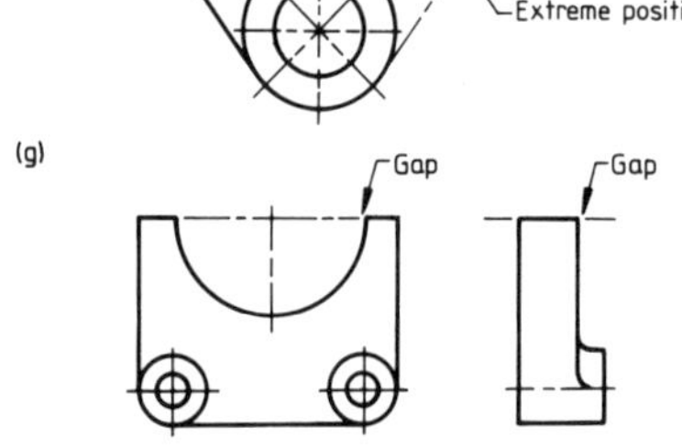

▲**Figure 1.13** Applications of centre lines

Centre lines

Centre lines are used to indicate centres of circles, axes of cylinders, cones and other parts with circular cross-sections, lines of symmetry, pitch circles and pitch lines, and paths of motion. They also show arc centres where these have to be positioned relative to another feature.

Centre lines are thin and are made up of alternate long and short dashes. The spacing between long and short dashes should be consistent, as should the length of the short dashes. The length of the long dashes depends on the length of the centre line; the longer the line the longer the dashes should be, up to a maximum of about 35 mm. Some CAD software produces centre lines in which the long dashes are only three or four times as long as the short dashes. This gives a drawing with long centre lines an amateurish and fussy appearance. Although in the illustration of the centre line in the table of line types in BS 308 : Part 1 the long dashes are quite short, this is dictated by the limited space available in the table and is not meant to represent typical practice.

Applications of centre lines are shown in Figure 1.13. At (a) a centre line defines the axis of the longitudinal view and two crossed centre lines show the centre of the circular view. This point is the end view of the axis. Note that centre lines should begin and end with a long dash and should cross at long dashes. However, some CAD systems may not allow these requirements to be easily controlled. Centre lines should extend a short distance only beyond the feature for which they are drawn. They should not continue through the space between views, except in rare cases where this will assist the dimensioning of a view.

The centre lines for small circles, below about 12 mm diameter, should be drawn with thin continuous lines, as in Figure 1.13(b).

A centre line is used for the line of symmetry of the left-hand view in Figure 1.13(c), for the pitch circle for the holes in Figure 1.13(d), for the pitch line for the holes in Figure 1.13(e) and for the path of motion in Figure 1.13(f). Always leave a gap, as in Figure 1.13(g), where the centre line forms a continuation of a visible or hidden line.

Do not show the centre lines of unimportant rounded or filleted corners.

Additional information

Where centre lines are the skeleton for a view they should be drawn first. Once they are all in place the view can be built up on them.

Precedence of coincident lines

Where two or more lines of different types coincide, visible lines take precedence over centre lines and hidden lines, and hidden lines take precedence over centre lines. Assignment 3 describes the treatment of hidden lines.

A POINT TO NOTE

Alternative and extreme positions of a part, such as that for the lever in Figure 1.13(f), are shown using thin, double-dashed, chain lines.

Additional information

The full alphabet of lines used on engineering drawings is illustrated and their applications described in BS 308 : Part 1. Only those in everyday use are referred to in this workbook.

Partial views

When a small feature must be illustrated but a complete view is unnecessary, a partial view may be used. An example is shown in Figure 1.14. The direction of viewing is indicated by an arrow and the view is identified with a title.

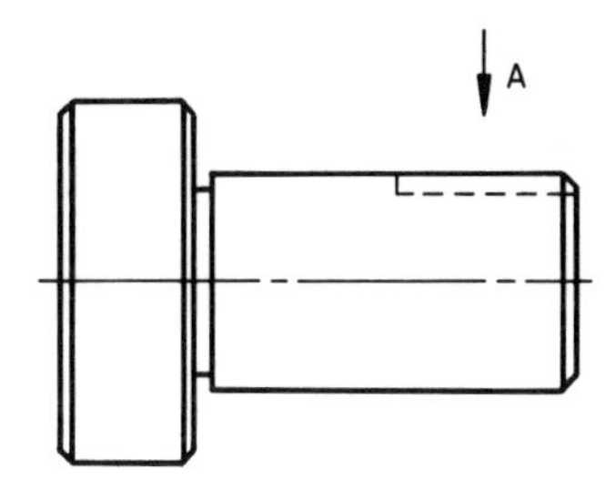

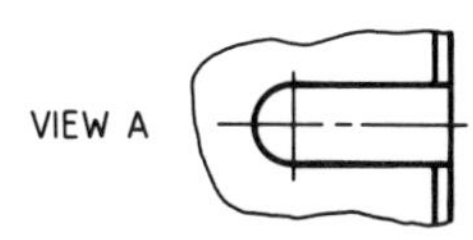

▲**Figure 1.14** Partial view

Symmetrical parts

Parts which are symmetrical about centre lines need not be drawn in full. Instead, a fraction of the whole view may be drawn, usually a half or a quarter. In Figure 1.15(a) symmetry symbols made up of two short thin parallel lines are drawn at each end of the lines of symmetry at right angles to them. The length of these lines should be about four times the distance between them.

Alternatively, the symmetry symbols may be omitted and the outlines of the part extended a short distance beyond the line of symmetry, as in Figure 1.15(b).

The symmetry convention may still be used even if the part has some asymmetrical features. A fractional view which shows the asymmetrical features must be drawn and a note added that identifies them. Figure 1.15(c) is an example.

Draw complete views if there is any possibility that the use of this convention will be misunderstood.

A POINT TO NOTE

The irregular boundary of a partial view is a thin line.

(a)

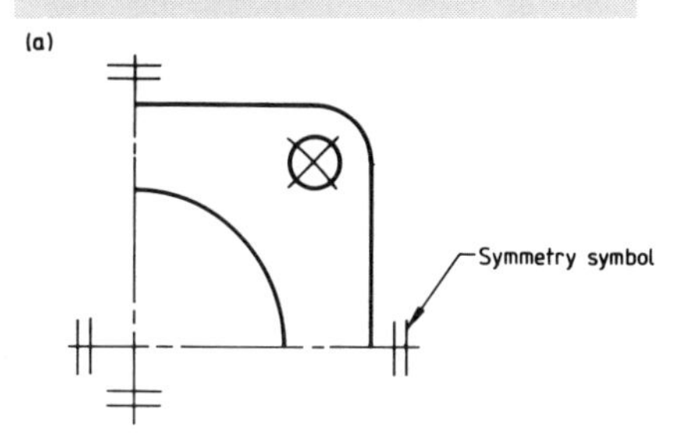

(b)

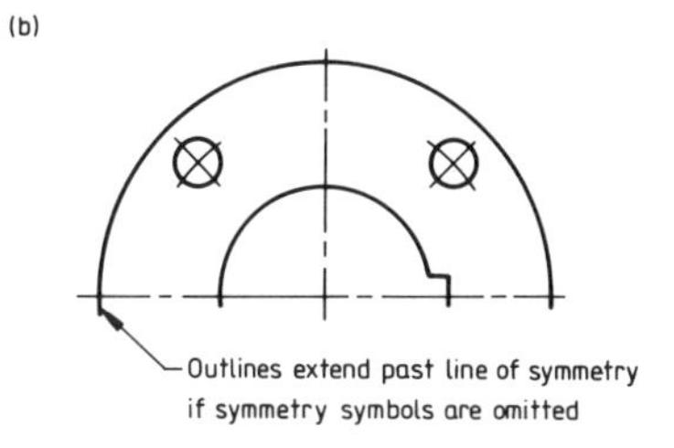

(c)

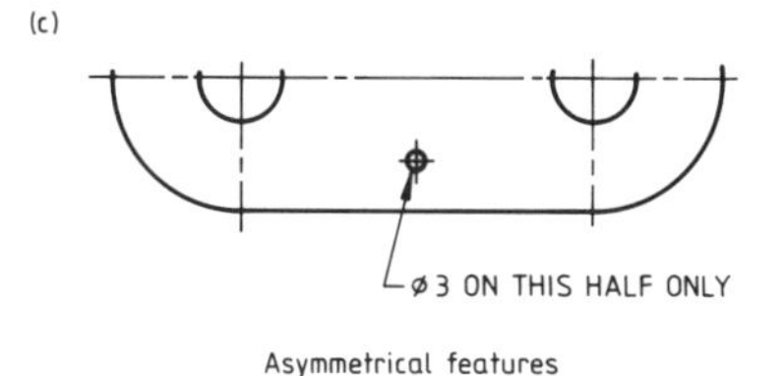

Asymmetrical features

▲**Figure 1.15** Symmetrical parts

Repetitive features and parts

Identical features and parts, such as holes, slots, and nuts and bolts which appear several times on a view need not all be drawn. Instead, one may be shown and the positions of the others indicated by their centre lines, as in Figure 1.16(a).

(a)

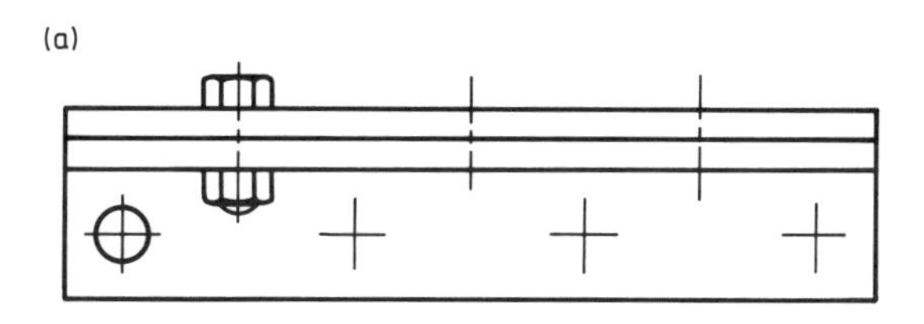

(b)

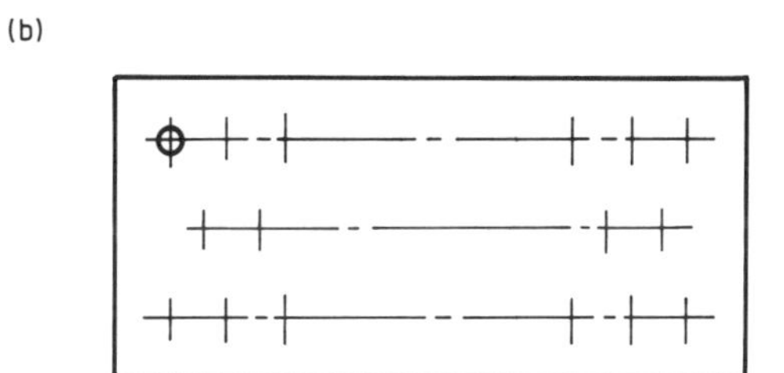

(c)

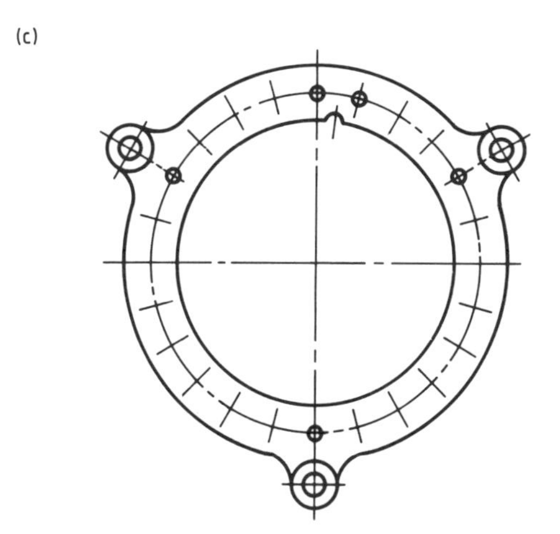

◀**Figure 1.16** Repetitive features and parts

Where a regular pattern of holes, bolts, rivets, slots and so on is required, as for the plate in Figure 1.16(b), one of them should be shown in full and sufficient of the remainder to establish the pattern should be shown by their centre lines. The remaining information should be given in a note.

Where a single feature is positioned near to and in relationship with one or more repetitive features, as in Figure 1.16(c), those repetitive features must be shown in full to establish the relationship.

TASKS

Task 1.3

Draw three suitable undimensioned views of the shaft support in Figure 1.17 using first-angle projection. Show all centre lines; hidden lines are not required.

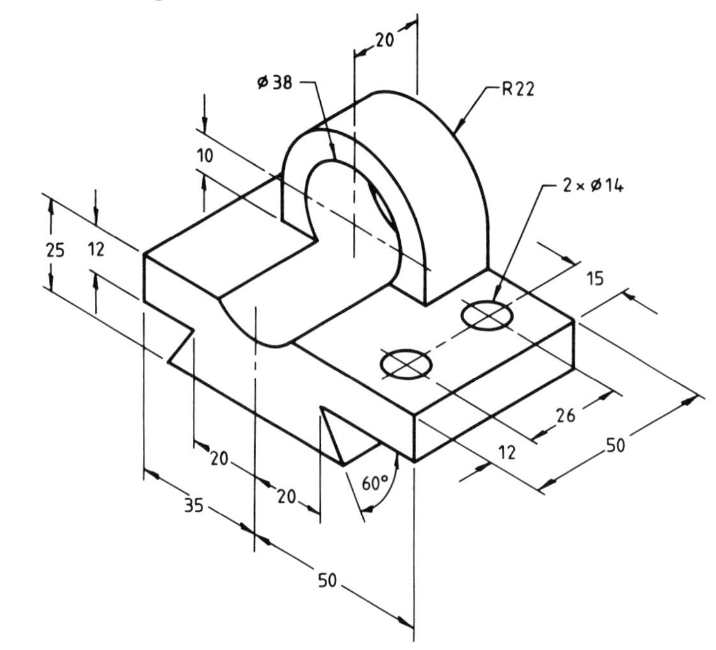

▶**Figure 1.17** Shaft support

Learning Assignment 2

Scales

The proportions of the drawing of an object must be the same as the proportions of the object itself. This means that the drawing must be made to scale. Objects which when drawn full size are too large for the largest available standard drawing sheet will need to be drawn smaller than full size. Conversely, it may be impossible to make full size drawings of very small objects and such parts must be drawn larger than full size.

BS 308 recommends that the scale which has been used is stated on the drawing as a ratio. An example would be ORIGINAL SCALE 1:1, which means full size. The scale should not be designated in words, such as 'twice full size' or 'half full size'.

Reduction scales

To make a drawing of an object smaller than full size a reduction scale is used. Reduction scales recommended in BS 308 which are in common use are 1:2, 1:5, 1:10, 1:20, 1:50 and 1:100.

Enlargement scales

These are used when the drawing has to be made larger than full size. Enlargement scales recommended in BS 308 are 2:1, 5:1, 10:1, 20:1 and 50:1.

Additional information

The term 'original scale' is used rather than merely 'scale', because drawings are frequently reproduced on larger or smaller sheets than those used when they were first drawn. Such reproductions are therefore at different scales from the original. The use of the term is a reminder not to try to take measurements from the print with a rule.

It is sometimes found that the gap between scales of 1:2 and 1:5, and 2:1 and 5:1 is too large. In this event scales of 1:4 and 4:1 are used, although they are not recommended in BS 308.

Choice of scales

The selected scale should be large enough to allow the drawing to be drawn, dimensioned and interpreted easily. If some details are too small for clear dimensioning at that scale, they should be shown in a separate view to a larger scale. This is usually a partial view. The main scale must then be indicated as the original scale and the scale used for the enlarged view must be shown close to it.

A POINT TO NOTE

Where tasks in this workbook require a drawing to be made, remember to choose an appropriate scale for it.

TASKS

Task 2.1

The instrument plate shown in Figure 2.1 is too small to be drawn full size. Using a suitable recommended scale draw the given view on a standard A4 drawing template. Do not show any dimensions, and position the drawing centrally on the sheet.

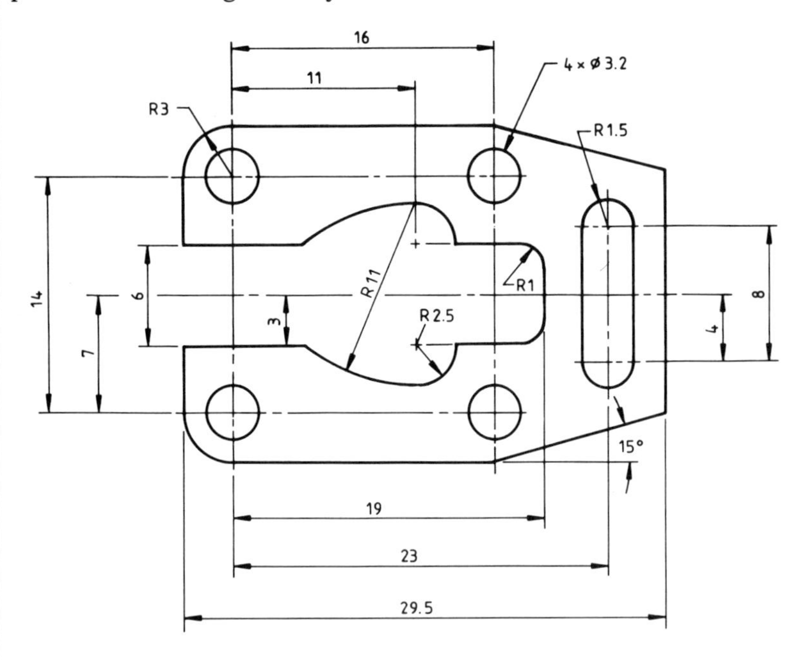

▶**Figure 2.1** Instrument plate

Task 2.2

Draw three appropriate orthographic views of the shaft bracket shown in Figure 2.2. Choose a recommended scale which will show how the drawing will appear when reduced to fit on a standard A4 drawing template. Dimensions and hidden lines are not to be shown.

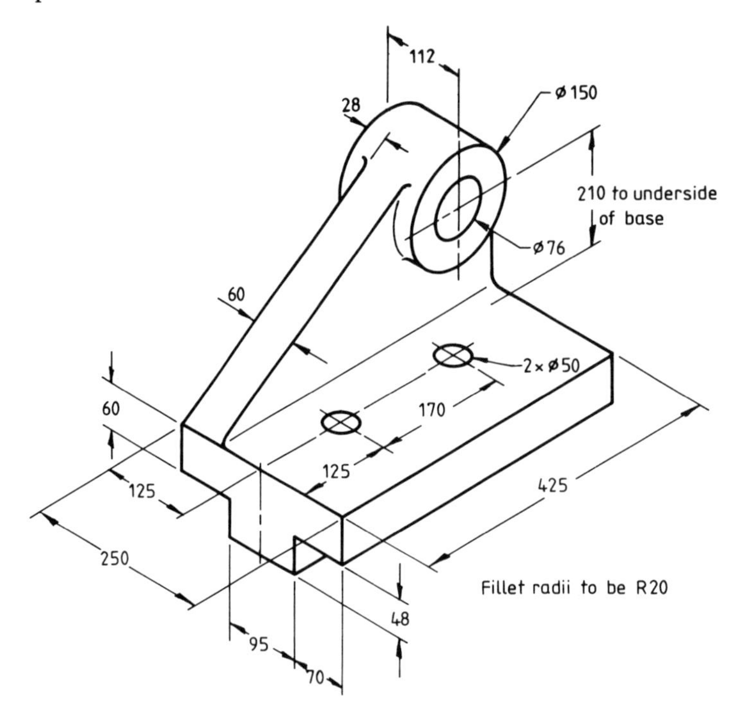

▶**Figure 2.2** Shaft bracket

Learning Assignment 3

Hidden detail

Engineering parts, almost without exception, have features which are internal and therefore cannot be seen in the selected orthographic views of the part. If it is necessary to show such hidden detail to describe the part completely, then thin dashed lines are used instead of the thick continuous lines used for visible features.

Everything which is visible in a view must be shown, but it is not necessary to show all hidden detail in every view. Often, if all of it were shown it would confuse the drawing rather than clarifying it. In addition, the drawing would take longer to read because every line must be understood, even if, on inspection, it proves to be unnecessary. Knowing how many hidden features to include on any drawing comes with experience, but if views are chosen in which as much information as possible is provided by visible lines then the need to add hidden detail will be reduced to a minimum.

Hidden lines are thin. The dashes and the gaps between them should be of consistent lengths, except for short lines when the length of the dashes may be reduced.

Figure 3.1 shows correct practices in drawing hidden lines.

1. Hidden lines touch visible lines and other hidden lines.
2. A gap is left where a hidden line dash is a continuation of a visible line.
3. Where a hidden line crosses a visible line it should preferably do so at a gap in the hidden line. This is not always possible.
4. Where two or three hidden lines meet at a point, the dashes should join. Note particularly how this is applied to the 120° conical end to a drilled hole, as illustrated in Figure 3.1(c).
5. Parallel hidden lines which are close together should be drawn so that the dashes are staggered.
6. Hidden arcs should start with a dash at tangent points.

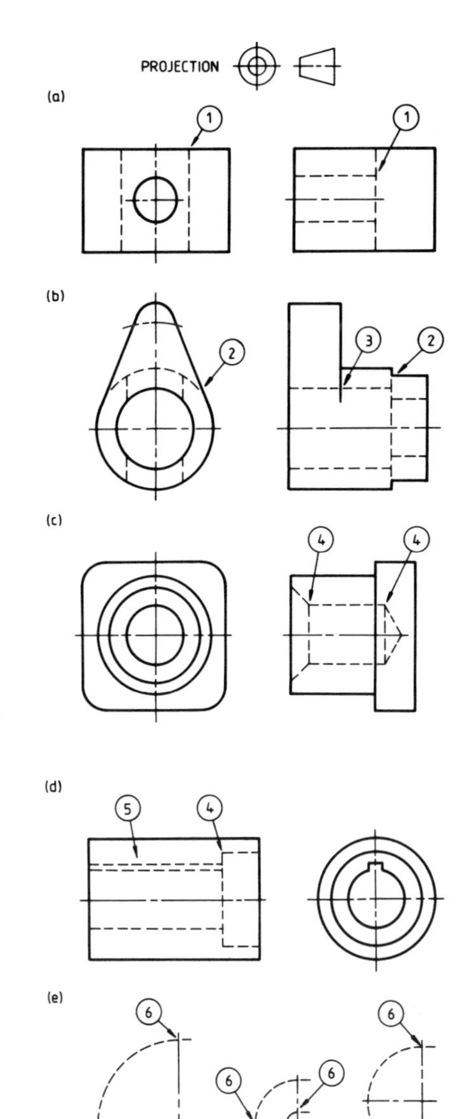

▲**Figure 3.1** Treatment of hidden lines

TASKS

Task 3.1

Draw a front elevation in the direction of arrow A and a plan in the direction of arrow B of the spacer in Figure 3.2. Use first-angle projection. All hidden detail is to be shown, but dimensions are not required.

Hole P is first drilled through at 10 mm diameter and then counterdrilled to 28 mm diameter for a depth of 25 mm. This is the

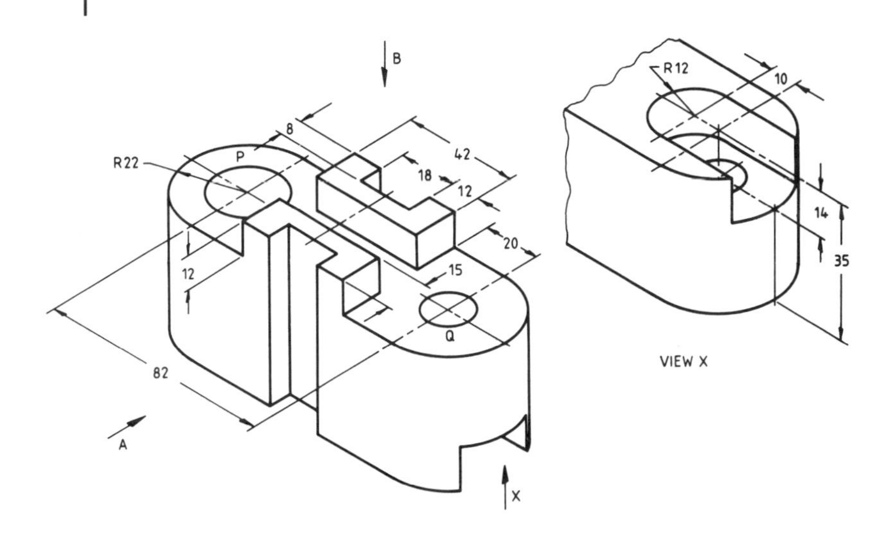

◀**Figure 3.2** Spacer

depth of the cylindrical part of the counterdrilling. Note that part of the conical end of the counterdrilling will be present.

Hole Q is a through hole 20 mm in diameter.

Task 3.2

Using third-angle projection, draw the following undimensioned views of the setting block in Figure 3.3:

- A front elevation in the direction of arrow A;
- A plan in the direction of arrow B;
- An end view in the direction of arrow C.

All hidden detail is to be shown in all three views.

Hole R is 18 mm in diameter and drilled through. The three bosses S are 18 mm in diameter and 8 mm high. They are drilled 10 mm in diameter by 12 mm deep.

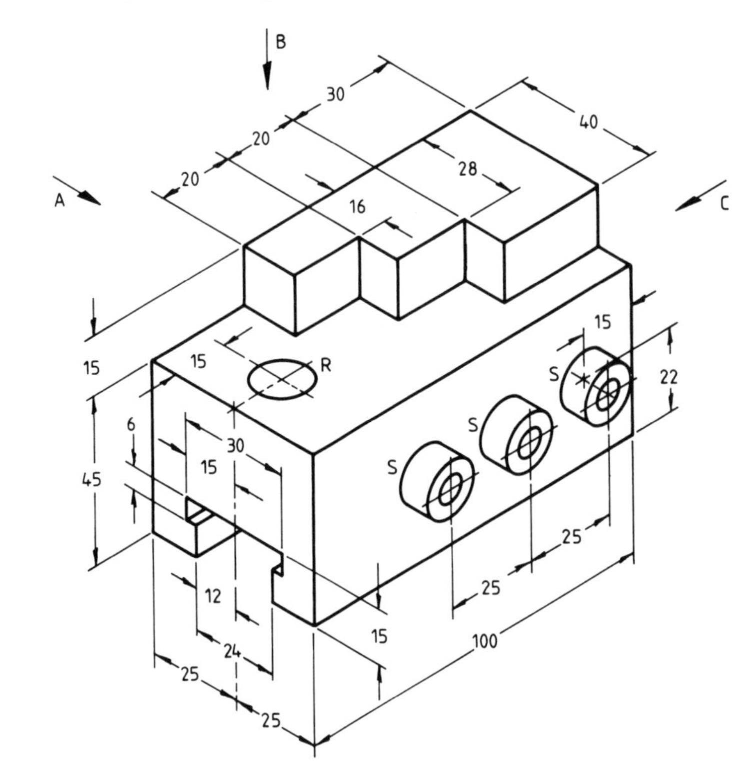

▶**Figure 3.3** Setting block

Learning Assignment 4

Sectional views

The method which is used to represent objects in orthographic projection was explained in Assignment 1. With this method a limited number of carefully chosen views will fully describe the exterior of the most complicated object.

However, it is very often necessary to draw objects having interior features, and these must be shown if the object is to be described completely. Interior features can be represented by hidden lines, as discussed in Assignment 3, but interpretation of hidden detail is difficult when the interior features are complicated. For clarity the interior construction needs to be shown using outlines. This can be achieved by replacing one or more of the outside views of the object by a sectional view.

To draw a sectional view, imagine the object to be sliced through by a plane, as shown in Figure 4.1(a). In Figure 4.1(b) the cutting plane has been removed, together with the part of the object between the observer's eye and the plane. Now the interior construction is exposed and can be drawn using outlines. The resulting sectional view is shown in Figure 4.1(c).

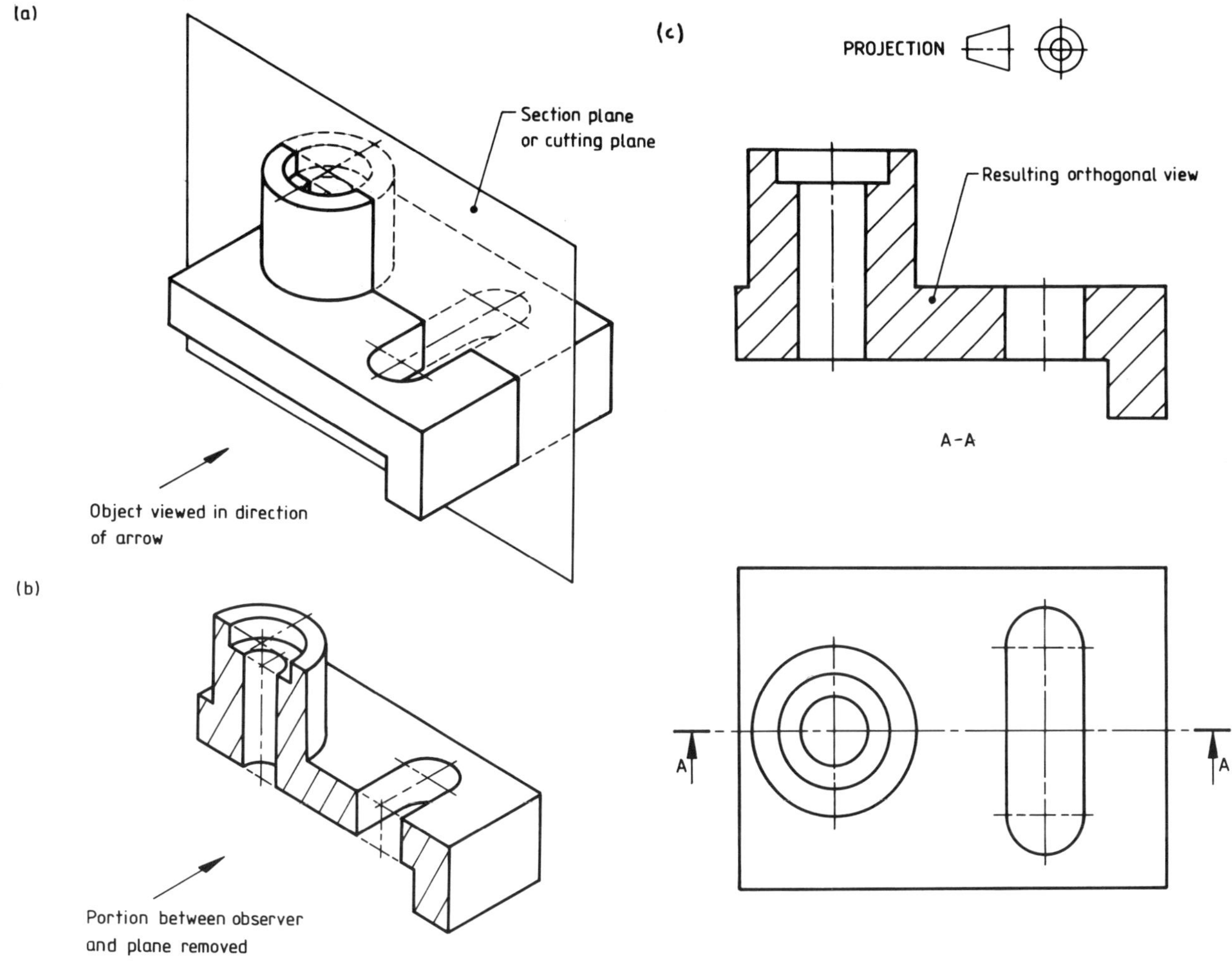

▲**Figure 4.1** Interior shown by a section

POINTS TO NOTE

- All visible lines behind the cutting plane must be shown, or the sectional view will be incomplete. See Figure 4.2.
- Hidden detail lines are not shown on a sectional view unless they are needed to describe the object completely. See Figure 4.3.
- The part of the object between the observer and the plane is only imagined to be removed. It is not actually shown removed anywhere on the drawing except on the sectional view itself. So the plan view in Figure 4.1 (c) must be a complete view.

The sectional view obtained by passing the cutting plane completely through the object is often called a full section.

To distinguish a sectional view from an outside view, cross-hatching may be drawn on the cut surfaces produced by the cutting plane. Hatching is discussed in detail in Assignment 5, but is shown in the figures illustrating this assignment for completeness.

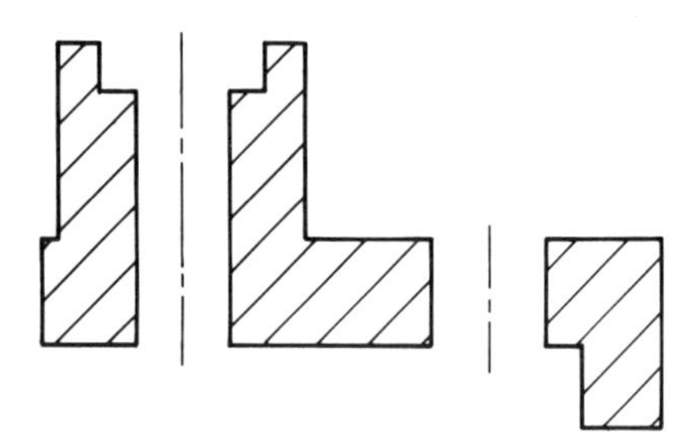

▲**Figure 4.2** Background lines on sectional views

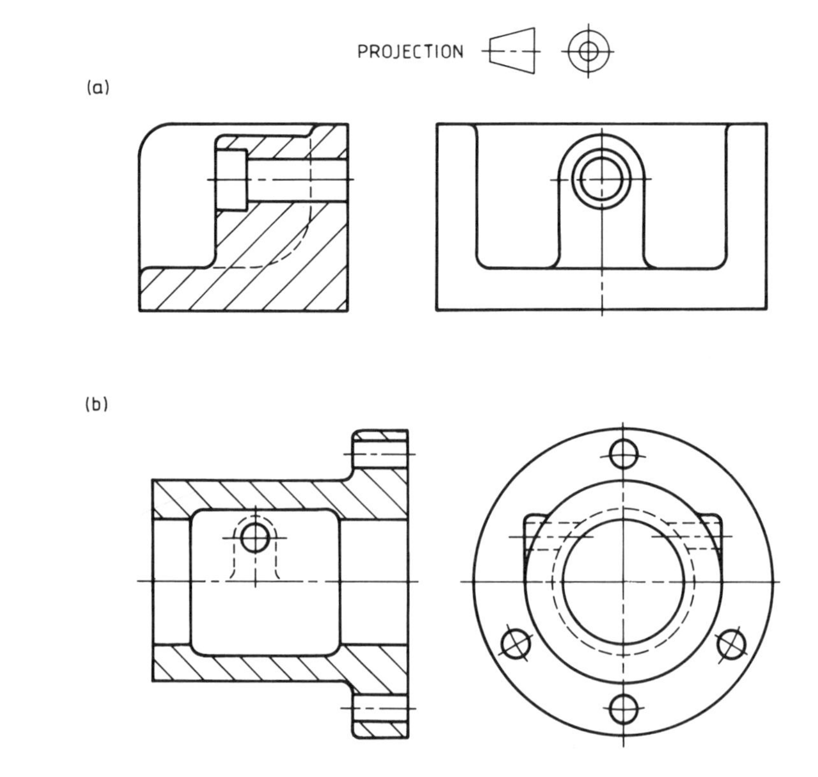

▶**Figure 4.3** Hidden detail added to sectional views for clarity

Cutting planes

The position of the cutting plane should be chosen to show the interior of the object to the best advantage. The position is shown as a line on a view adjacent to the sectional view. The cutting plane line is a thin, long-chain line with a thick, long dash at each end. The proportions of the parts of the line should be as shown in Figure 4.1 (c).

The direction in which the cutting plane is to be viewed is shown by arrows at the ends of the line. These are placed with their points at the centres of the long dashes. The arrowheads should be 7 mm to 10 mm long, with the stem of the arrow a little longer than the arrowhead. The cutting plane is identified by capital letters and the sectional view is related to the plane by a title such as 'A–A', placed below the view. A different letter must be used for each cutting plane on the drawing.

The cutting plane and the sectional view title may be omitted when the position of the plane is obvious. One example would be where the plane passes through the centre line of a symmetrical view of the object. This is the case in Figure 4.1 (c), and the plane and title are shown there for illustration only.

POINTS TO NOTE

- When the arrowheads are drawn by a CAD machine they need not be filled in. In this case they should be drawn with thin lines.
- BS 308 recommends that the letters identifying the cutting plane are placed near the stems of the arrows. Occasionally lack of space prevents this being done.
- In general there is no need for the word 'SECTION' to appear in the title of the sectional view.

Additional information

It is sometimes convenient to change the direction of the cutting plane as it passes through the object. Sectional views where this is done are illustrated and explained in BS 308 : Part 1. At each change of direction thick dashes should be used, of the same length as those at the ends of the plane.

In the following tasks show and identify the cutting planes, but do not add the cross-hatching. Retain your solutions so that the hatching can be added when Assignment 5 has been completed.

Task 4.1

Replace the given front view of the bearing block shown in Figure 4.4 with section A–A and project from it section B–B. Do not dimension your solution.

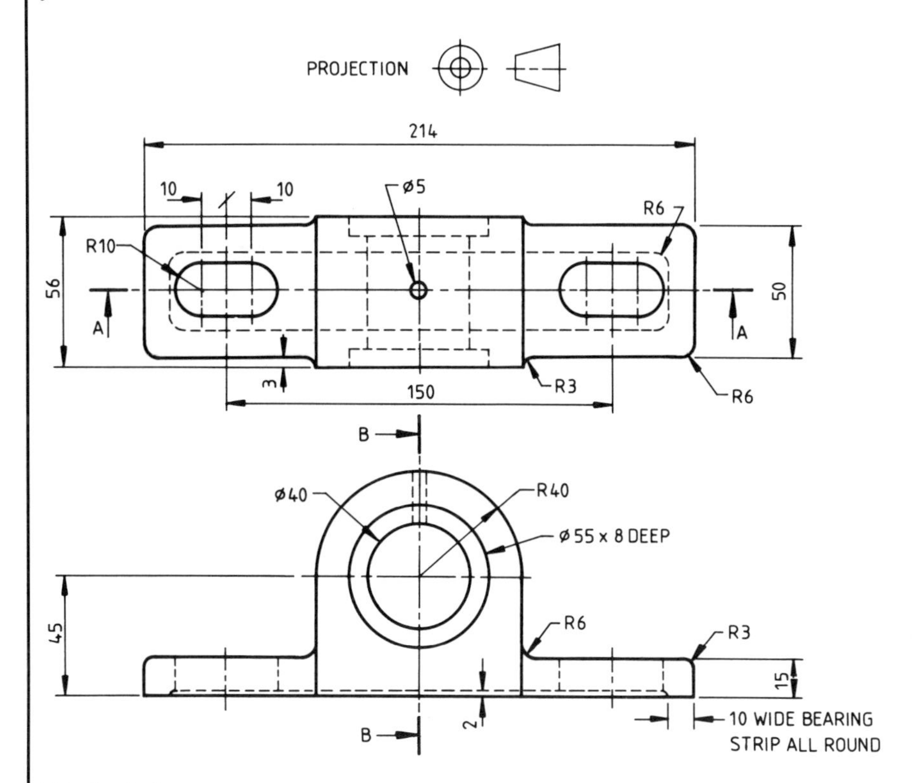

◀**Figure 4.4** Bearing block

Task 4.2

Draw the given left-hand view of the cover in Figure 4.5 and project from it section A–A and section B–B. Do not show any dimensions.

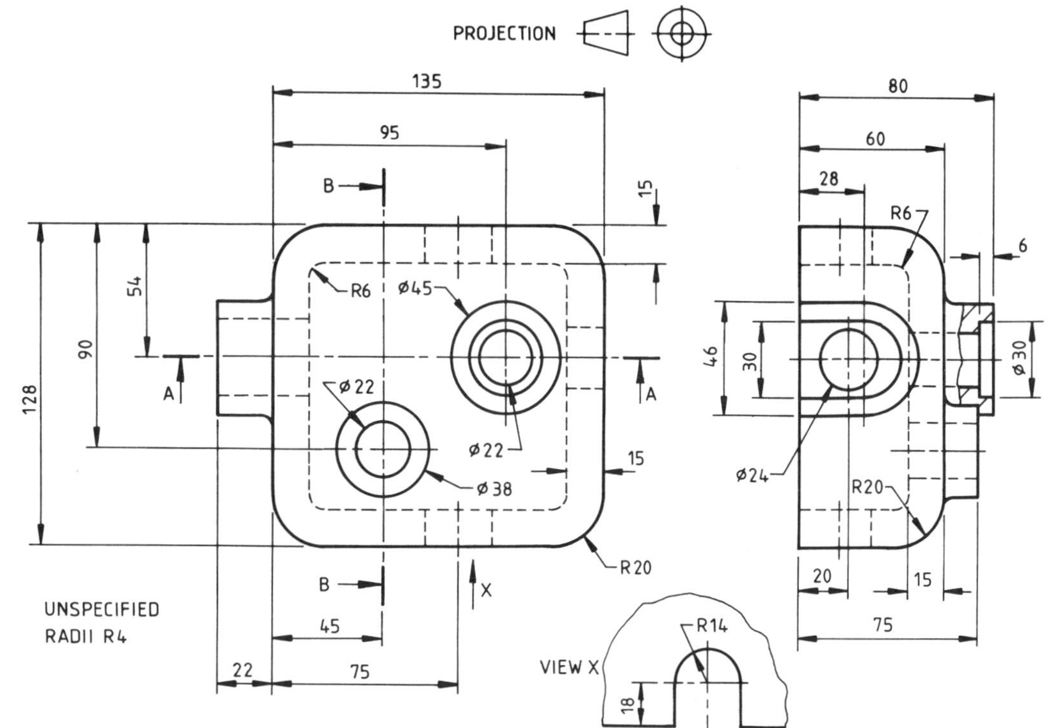

◀**Figure 4.5** Cover

Learning Assignment 5

Hatching or section lining

As noted in Assignment 4, to distinguish a sectional view from an outside view section lining or hatching may be drawn on the surfaces produced by the cutting plane, as shown in Figure 4.1. Hatching uses thin lines inclined at 45°, which are equally spaced with a minimum distance between them of about 4 mm. The larger the area to be hatched the wider the spacing may be, up to about 10 mm. For very large areas, hatching may be reduced to a band around the edge of the hatched area, as shown in Figure 5.1. Note that hatching must touch the outline of the sectional view.

A POINT TO NOTE

An outline is never crossed by hatching.

The hatching on a section through a single component must have the same slope and spacing throughout the view, as illustrated in Figure 5.2(a). Figure 5.2(b) shows that this applies also to separated sectioned areas of the part.

On adjacent parts on an assembly drawing, the slope of the hatching should normally be reversed and staggered, as shown in Figure 5.3. For each part the slope and spacing of the hatching must be the same on all views of the drawing. When more than two parts touch, as in Figure 5.4, the hatching on two of them will slope in the same direction. In this event the lines must be staggered and are usually spaced more closely in the smaller hatched area.

A POINT TO NOTE

Only the simple hatching described above should be used on engineering drawings. Different materials should not be indicated by different types of hatching, such as dashed lines or alternate dashed and full lines.

The wide selection of hatching types available with CAD systems is useful when preparing construction drawings. These require different standardized hatchings for sections through brickwork, blockwork, stonework and subsoil, for example.

Additional information

It is becoming common to omit hatching from sectional views where the time and expense in drawing it cannot be justified. This practice is sanctioned by a recommendation in BS 308. However, it should only be used when there is no possibility of the view being misunderstood, otherwise time and money are likely to be wasted.

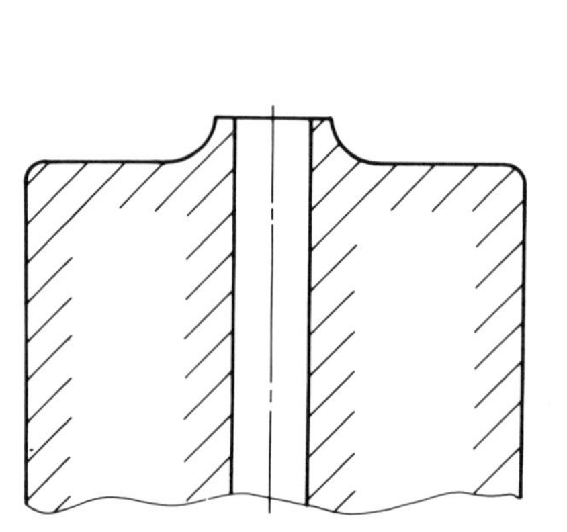

▲**Figure 5.1** Hatching large areas

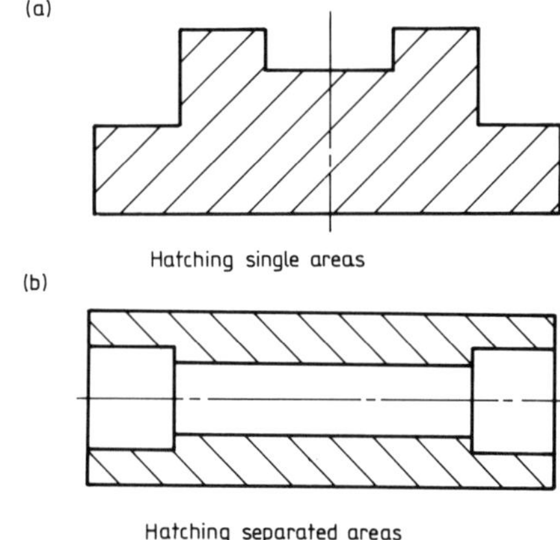

▲**Figure 5.2** Hatching on a single component

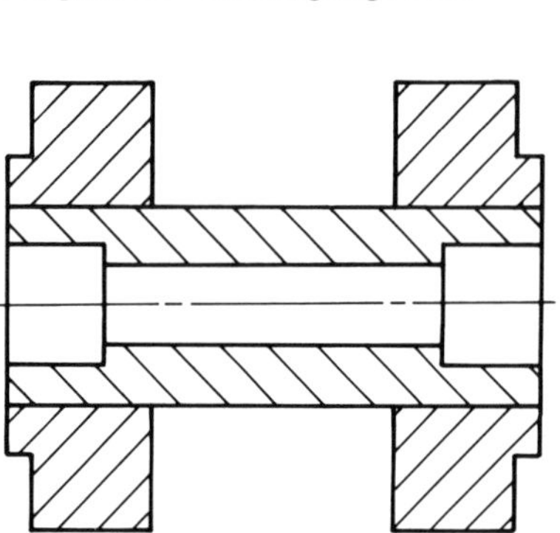

▲**Figure 5.3** Hatching adjacent parts

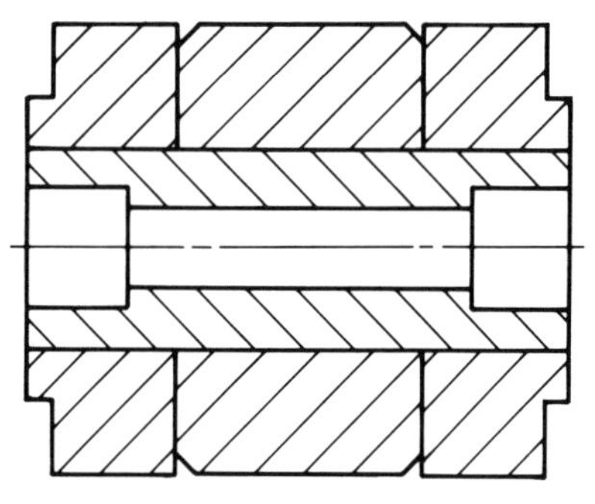

▲**Figure 5.4** Hatching adjacent parts at same angle

Task 5.1

Add the hatching to your solutions to Tasks 4.1 and 4.2

Task 5.2

Figure 5.5 shows the parts for a built-up roller. There are two end caps, which are pressed into the centre, and two bushes, which are pressed into the end caps. The flanges of the bushes are to be against the flanges of the end caps. Assemble the parts and draw a section through the centre line, using the correct representation for the hatching. Do not show any dimensions.

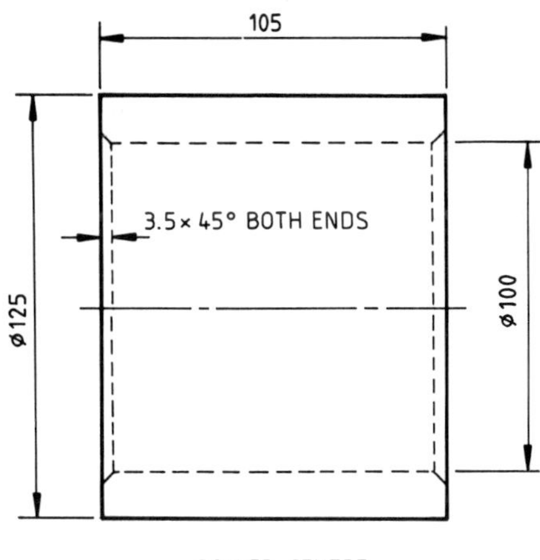

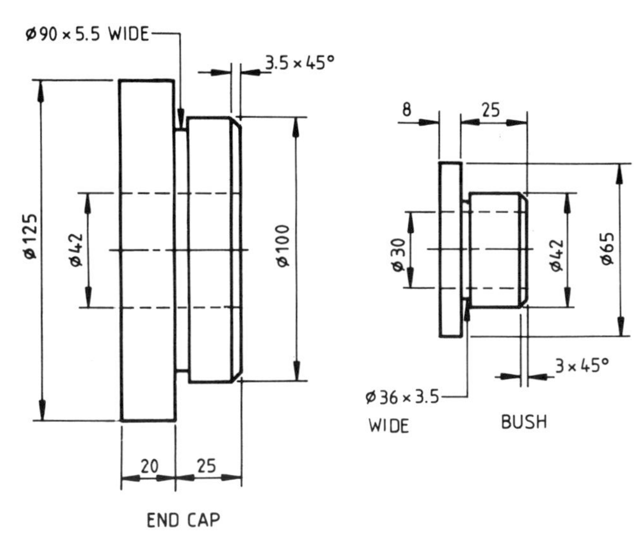

◀**Figure 5.5** Built-up roller assembly

END CAP

Alteration to the slope of hatching

Occasionally the outlines or axes of a sectional view slope at 45°. If the hatching also slopes at 45° then the view looks peculiar, as shown in Figure 5.6(a). The hatching should therefore be drawn at some other angle, such as 30° or 60°, or drawn horizontally or vertically. See Figure 5.6(b).

▼**Figure 5.6** Alteration to the slope of hatching

(a)

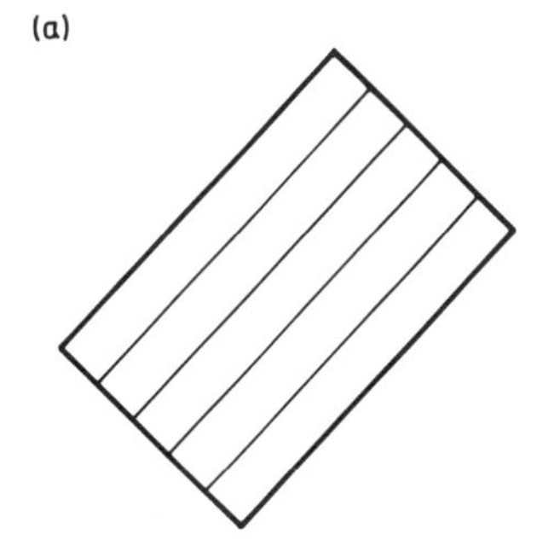

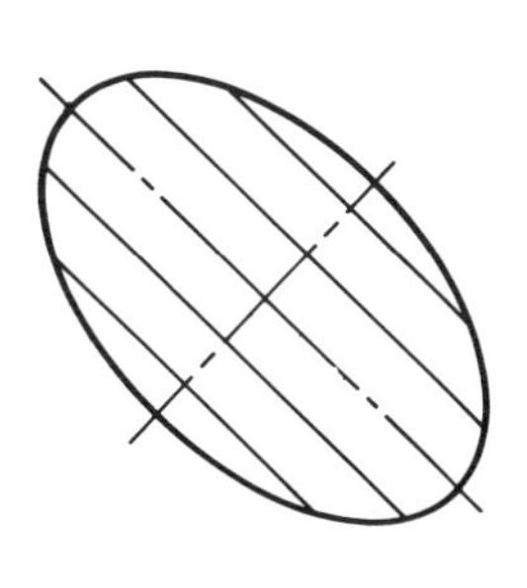

(b)

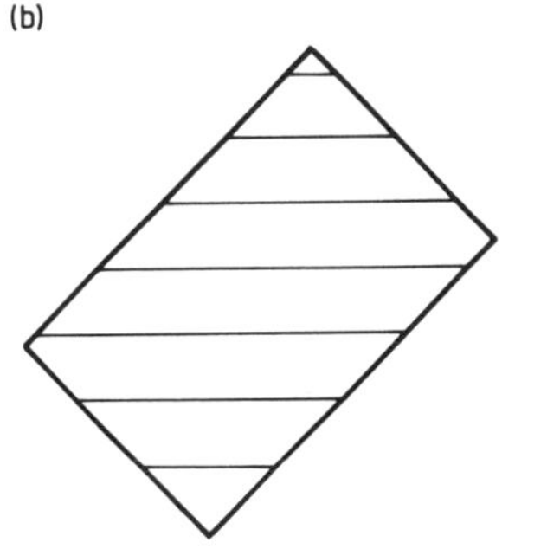

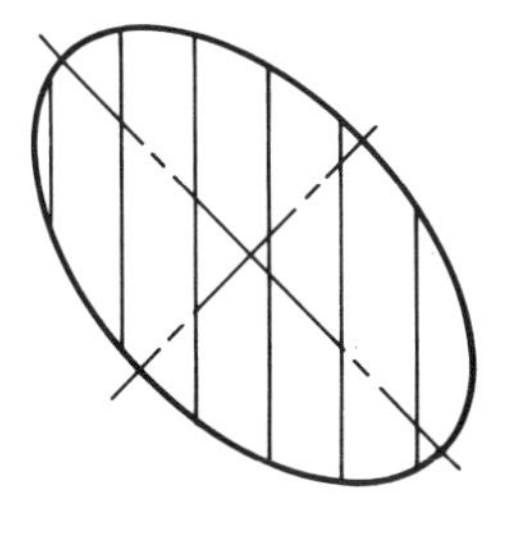

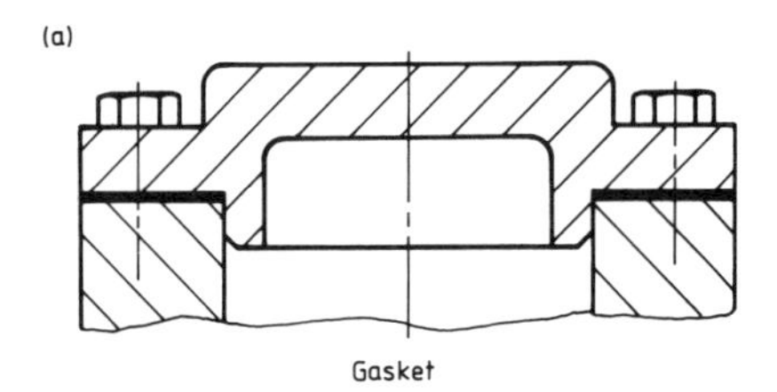

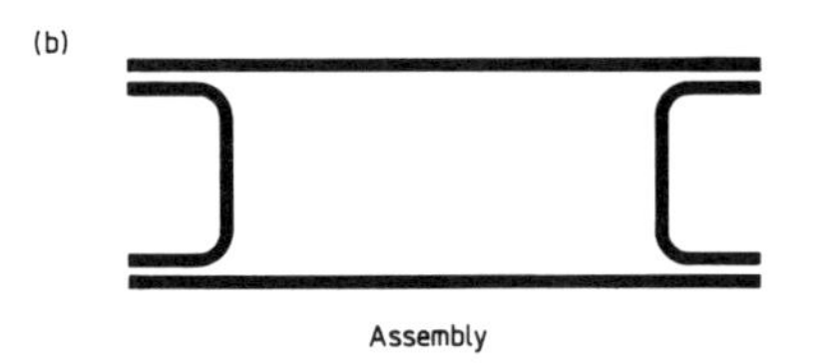

▲**Figure 5.7** Thin material in section

A POINT TO NOTE

The practice of drawing the thickness of the material out of scale, so that it can be hatched in the usual way, should be avoided.

Thin material in section

When gaskets and parts made from sheet metal and other thin material are cut by a section plane, the sectioned areas may be too narrow for hatching to be applied to them. In such cases the material thickness may be filled in rather than being hatched, as illustrated in Figure 5.7(a).

Figure 5.7(b) shows that adjacent thin parts in section should be separated by spaces to make the view easier to read. The spaces should not be narrower than 1 mm, or they may close up on reduced-size prints.

Features from which hatching is omitted

Whenever hatching on a sectional view could mislead the reader of the drawing it should be omitted. In Figure 5.8 section A–A cuts lengthwise through the triangular strengthening web on the part. If the web is hatched, as in Figure 5.8(a), then the part appears to be of constant thickness. To prevent this wrong interpretation, the hatching is omitted from the web, as in Figure 5.8(b). Note however that when the web is cut across by the section plane, as with section B–B in Figure 5.8(c), no misleading impression is given by the hatching and it must be shown.

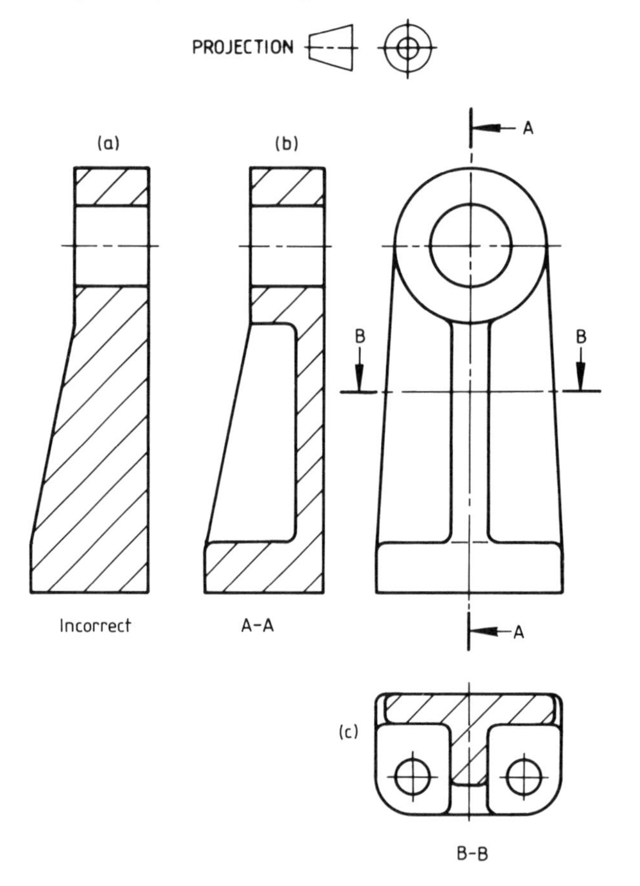

A POINT TO NOTE

The boundary of the web where it runs into the boss and base of the part in Figure 5.8(b) must not be left out. This boundary is shown as an outline, not a hidden detail line.

▶**Figure 5.8** Omission of hatching

Two more examples of the omission of hatching from sectioned features are shown in Figure 5.9. If section A–A in Figure 5.9(a) is drawn in true projection then the right-hand half of the part appears to be solid. In addition, the projections of the left-hand rib and the right-hand boss are awkward and confusing. Therefore, for clarity and ease of drawing, the conventional representation of section A–A should be drawn with the ribs, bosses and holes in the base imagined to be aligned on to the cutting plane.

Section X–X through the handwheel in Figure 5.9(b) should be treated in the same way. Spoke C should be revolved onto the cutting plane and, like spoke A, should not be cross-hatched. Spoke B should be omitted to avoid an awkward, time-consuming and confusing projection.

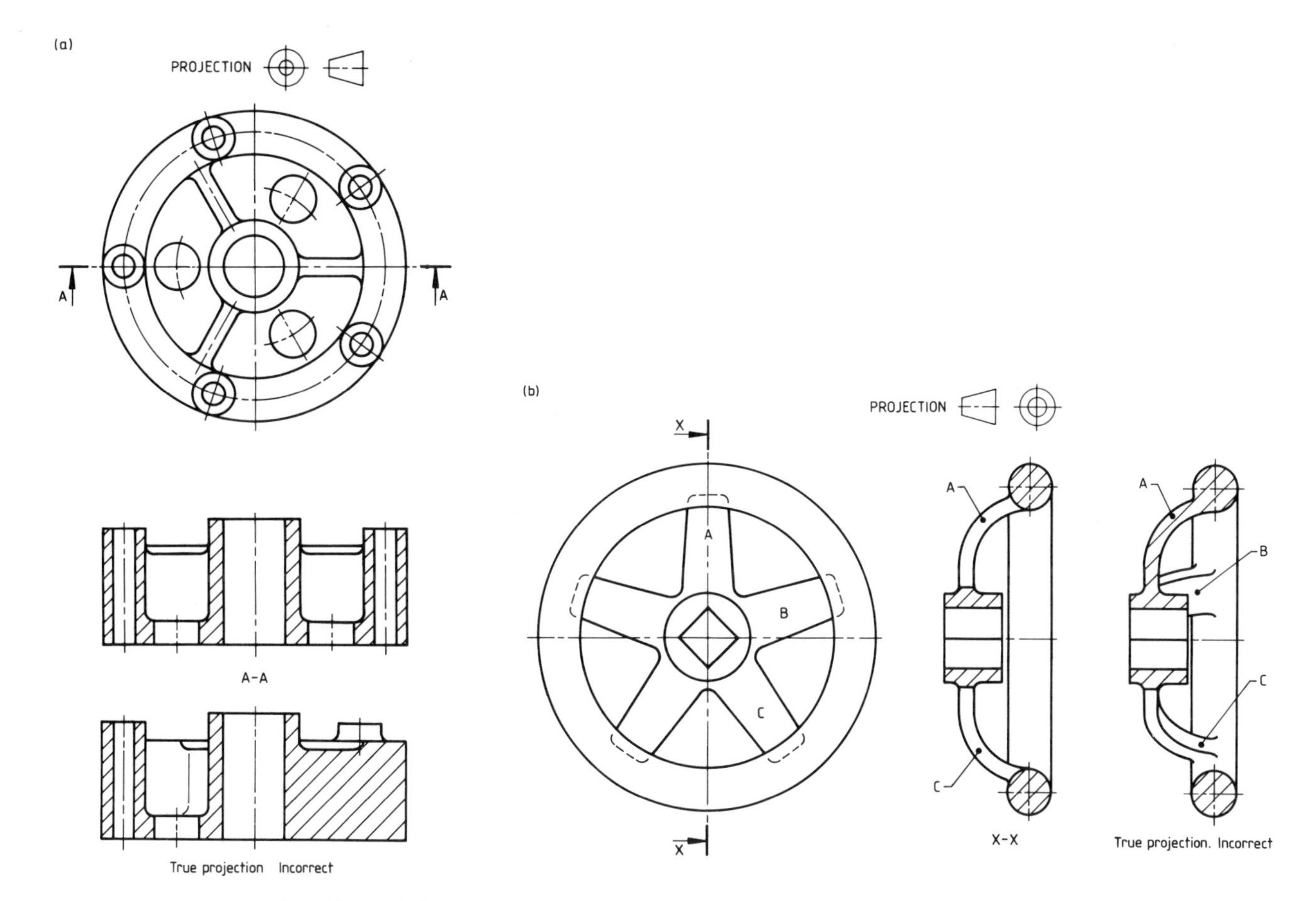

▲**Figure 5.9** Omission of hatching and alignment of features

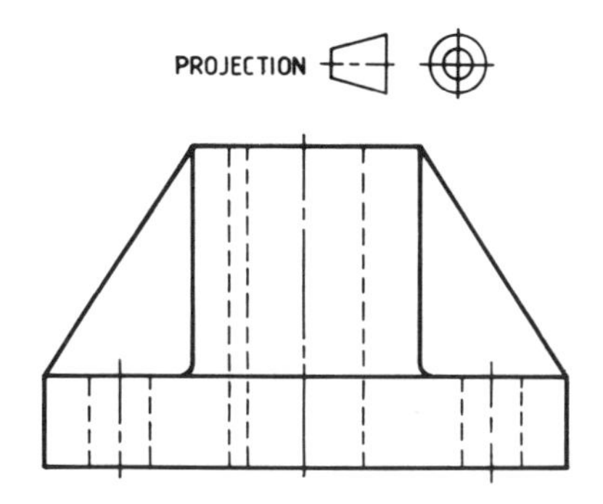

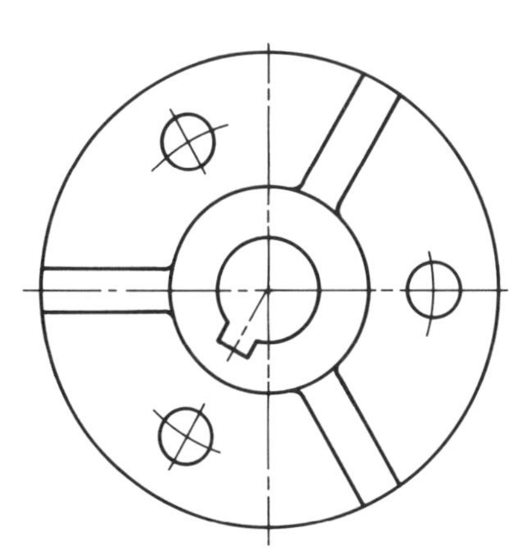

▲**Figure 5.10** Alignment applied to an outside view

Additional information

The alignment convention can be used on outside views as well as sections. In Figure 5.10, for example, the ribs, holes and keyway in the plan should be revolved to the horizontal centre line before being projected to an unsectioned front view. This will then be quicker to draw and easier to read than a view in true projection.

Parts which are not sectioned

Some parts are not sectioned when cut lengthwise by a section plane. Examples are given in Figure 5.11. They include nuts, bolts, screws, studs, washers, rivets, small solid cylindrical parts and solid shafts, keys, split and taper pins, and balls and rollers in rolling bearings. They are not sectioned because they have no internal features, and are more easily recognised by their outside views than by a section. However, if they are cut across by a section plane, so producing a circular sectional view such as A–A in Figure 5.11, then they are hatched in the normal way.

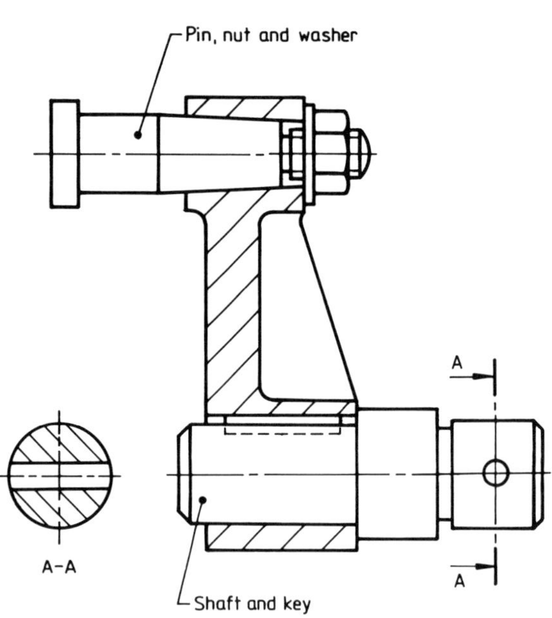

▶**Figure 5.11** Parts left unsectioned

TASKS

Task 5.3

Figure 5.12 shows the parts for a pivot pin assembly. The pivot pin fits in the left-hand bore of the base plate from above and is secured with the locking pin. Draw sections A–A and B–B of the assembled parts using the correct conventions and representation for the hatching. Dimensions are not to be shown.

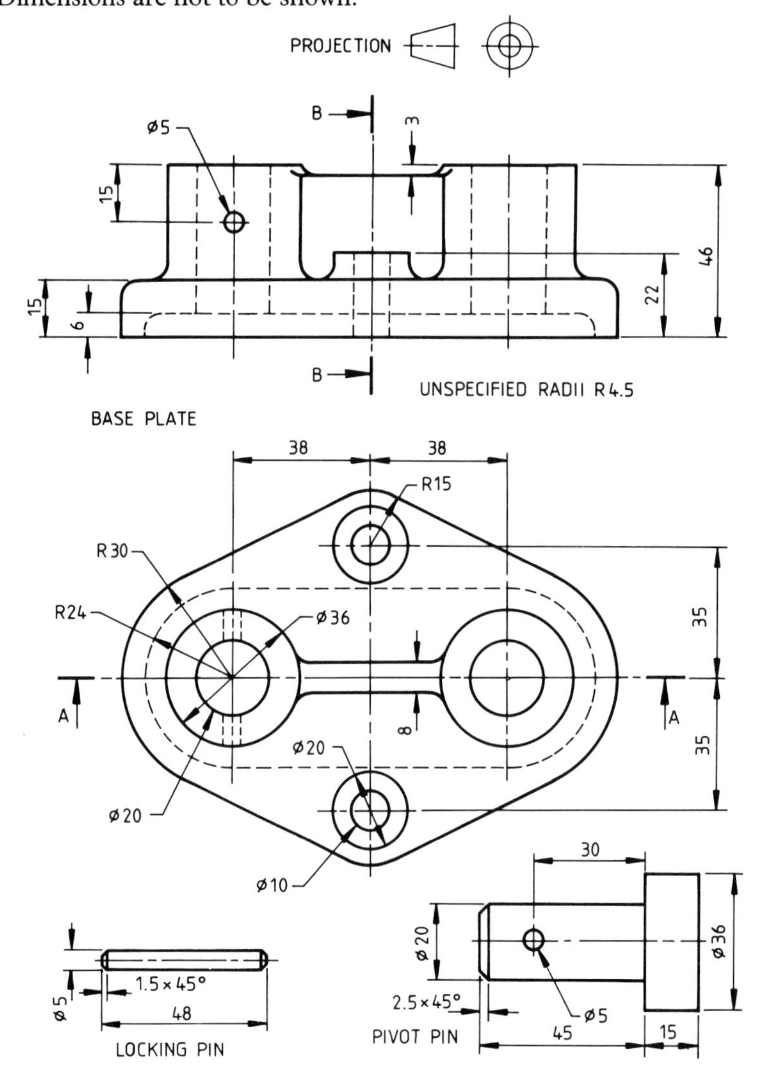

▶**Figure 5.12** Pivot pin assembly

Learning Assignment 6

Half sections, local sections, revolved sections and removed sections

Half sections

POINTS TO NOTE

- **The halves of a half section are separated by a centre line, not an outline.**
- **It is rarely necessary to identify the cutting plane or the view.**

It is not necessary for the cutting plane of a sectional view to pass completely through the object. If it is passed halfway through a symmetrical part then a half section results. One half of the view shows the exterior of the part and the other the interior. Thus two views are combined in one and draughting time and space on the drawing are saved.

A disadvantage of half sections of single parts, such as that in Figure 6.1 (a), is that if the interior features are relatively complex, then it may be difficult to dimension them without using hidden detail lines. In this event it is best to draw a full section. This problem does not arise with assembly drawings, because they need few, if any, dimensions, and hidden detail can therefore be omitted from the unsectioned half of the view. Figure 6.1 (b) shows a half section through an assembly.

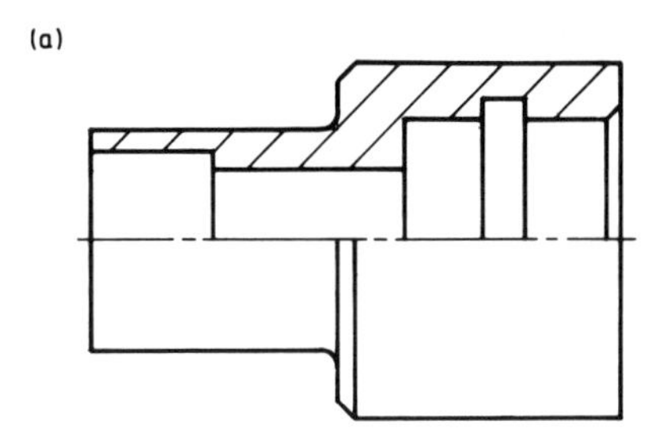

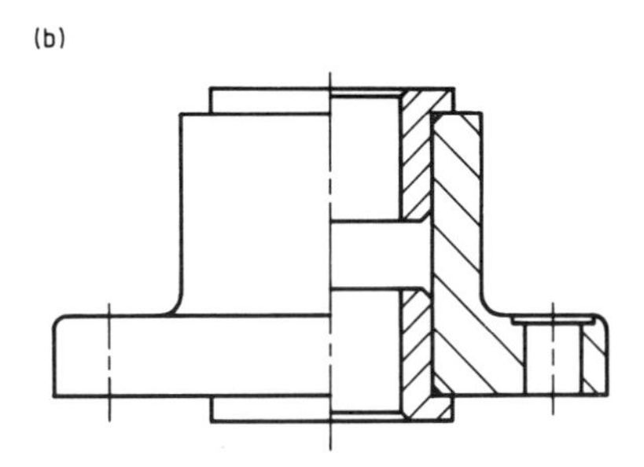

▲Figure 6.1 Half sections

TASKS

Task 6.1

Draw the left-hand view of the sealing cap in Figure 6.2, change the right-hand view to a half section and add a plan in half section projected from the left-hand view. Do not show any dimensions.

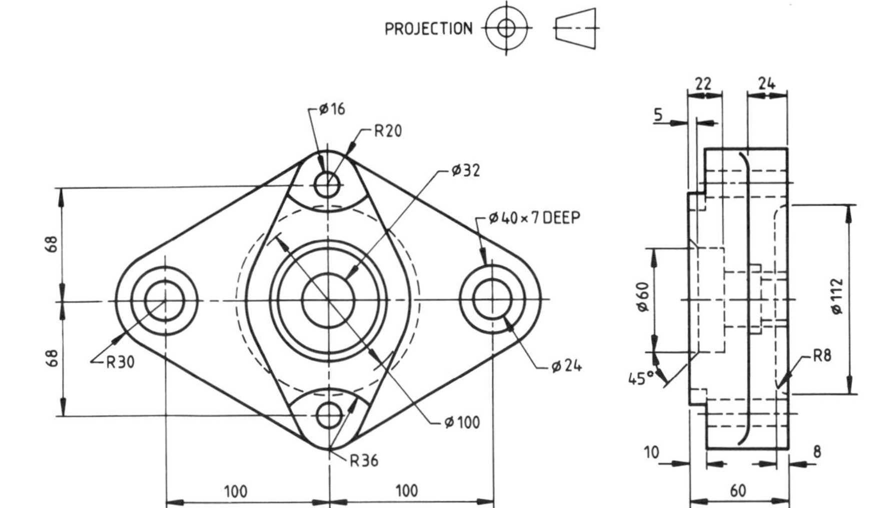

◀Figure 6.2 Sealing cap

Local sections

It often happens that only part of a view in section is needed to show interior features. In these cases a local section is useful. The cutting plane is passed part way through the component and is then assumed to be broken away, leaving an irregular boundary or break line. For this reason local sections are sometimes called broken-out sections. Examples of local sections are shown in Figure 6.3.

A POINT TO NOTE

The break line for a local section is thin.

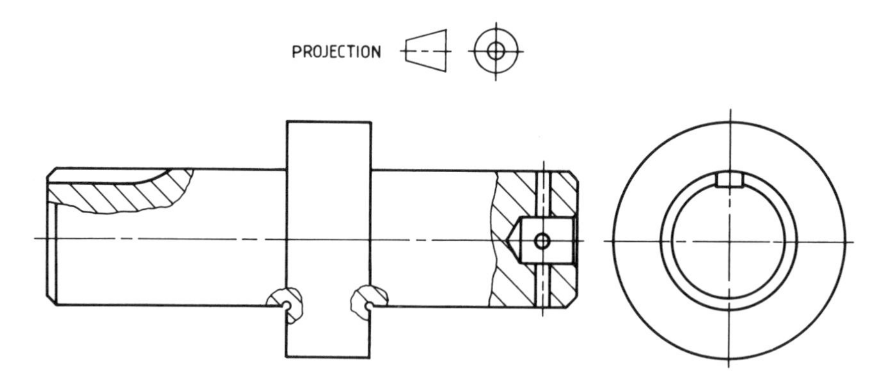

▶**Figure 6.3** Local sections

Additional information

The small undercuts on each side of the flange in Figure 6.3 enable the diameters to be ground for their full length without the grinding wheel fouling the sides of the flange.

POINTS TO NOTE

- The outline of a revolved section is a thin line.
- Even though hatching may be omitted in general from sectional views, in the interests of clarity it is usually best to add it to revolved sections if the main view is not broken to take the section. See, for example, the square section at the left of Figure 6.4(b).
- Where convenient, partial revolved sections may be used.

Revolved sections

These sections are used to show the local cross-section of an arm, rib or similar feature, and are drawn directly on an outside view. The cutting plane is assumed to pass through the feature at right-angles to its axis and then to be revolved into the plane of the paper. Figure 6.4 gives illustrations of revolved sections.

Figure 6.4(a) shows that any outlines on the main view which are covered by the section are omitted. The part in Figure 6.4(b) can have the variations in its cross-section simply depicted by the use of revolved sections. For clarity or ease of dimensioning, the section may be placed in a break in the main view, as shown in Figure 6.4(c). Figure 6.4(d) illustrates a common error in drawing revolved sections.

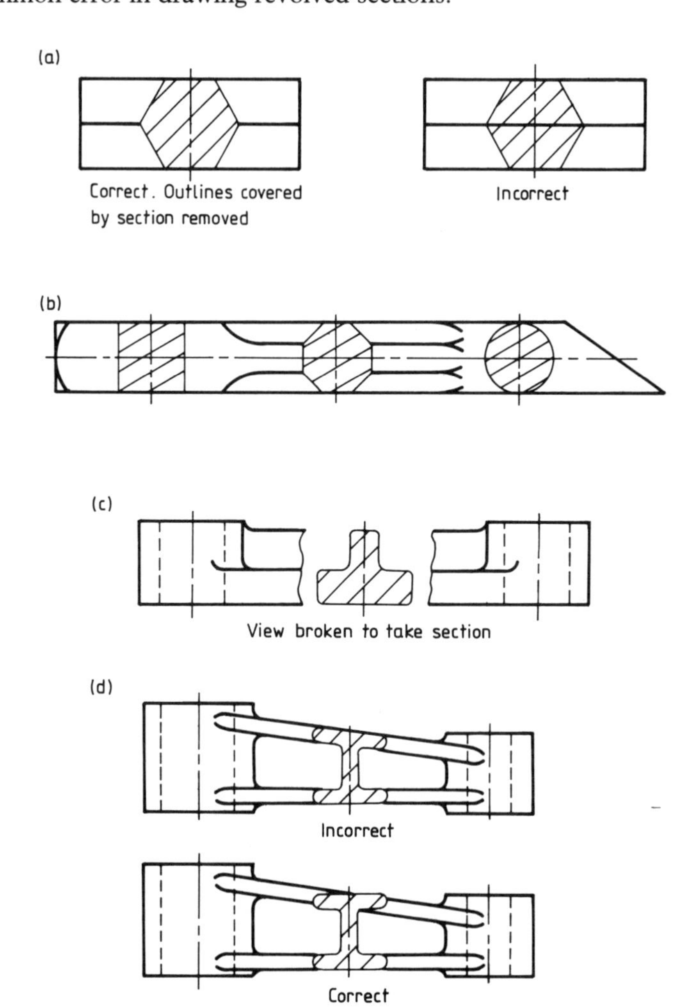

▶**Figure 6.4** Revolved sections

Removed sections

These are similar to revolved sections, but are positioned away from the main view instead of being drawn on it. They are used instead of revolved sections where there is not enough space for the section to be drawn or dimensioned on the outside view.

Removed sections need not be shown in their correct projected positions, but may be placed on the drawing wherever is most convenient. See Figure 6.5(a). If they are numerous, they are sometimes grouped together on a separate sheet. In either case, the position of the cutting plane must be shown, and the section must have a title to relate it to the cutting plane.

It is sometimes convenient to place removed sections near the main view on centre lines extended from the cutting planes. This is illustrated in Figure 6.5(b). Here there is no need to label either the cutting planes or the sections.

In Figure 6.5(c) successive removed sections are shown in order in their correct projected positions. With this arrangement the cutting planes and sections must be identified. Wherever possible the direction of viewing should be the same for all the sections.

POINTS TO NOTE

- The outlines of removed sections are thick.
- In general only features on the cutting plane are shown, features behind the cutting plane usually being left out, as recommended by BS 308. However, there are occasions when most people would show some detail behind the cutting plane. Examples are sections B–B and C–C in Figure 6.6.
- Removed sections are generally fairly simple, and hatching can often be omitted without making them difficult to understand.

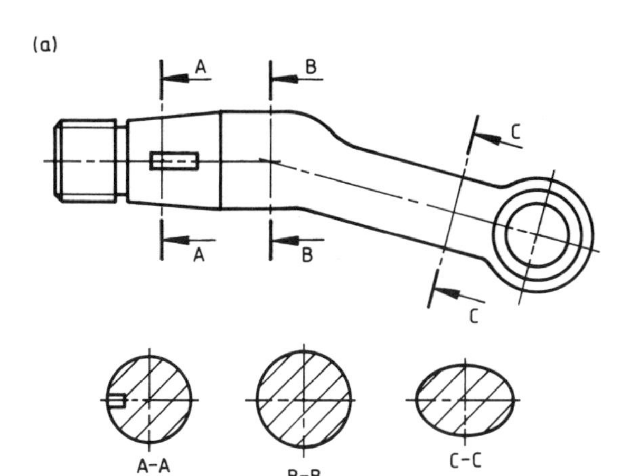

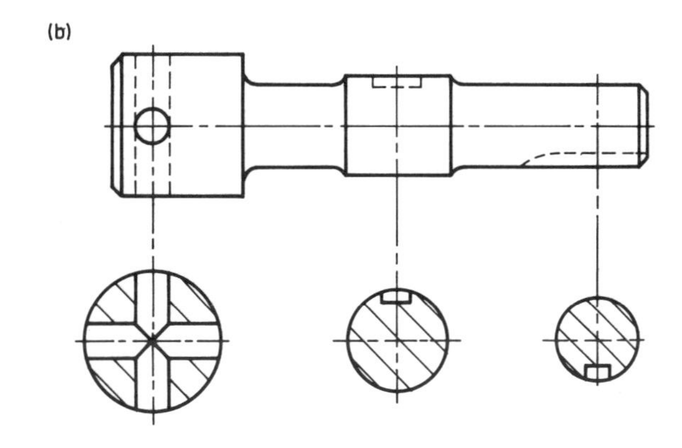

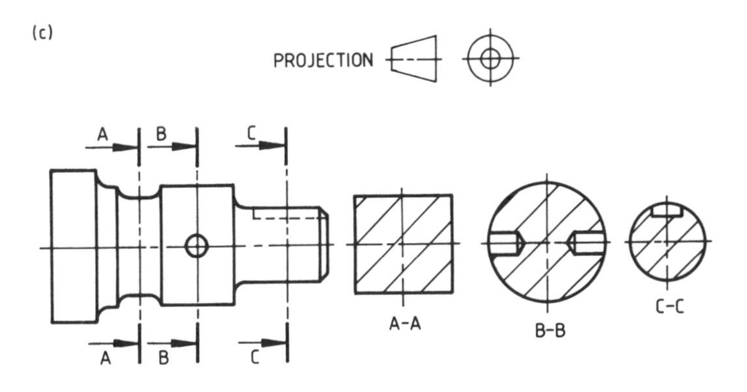

◀**Figure 6.5** Removed sections

TASKS

Task 6.2

Draw the given main view of the operating valve shown in Figure 6.6 and add the four removed sections. Dimensions are not required.

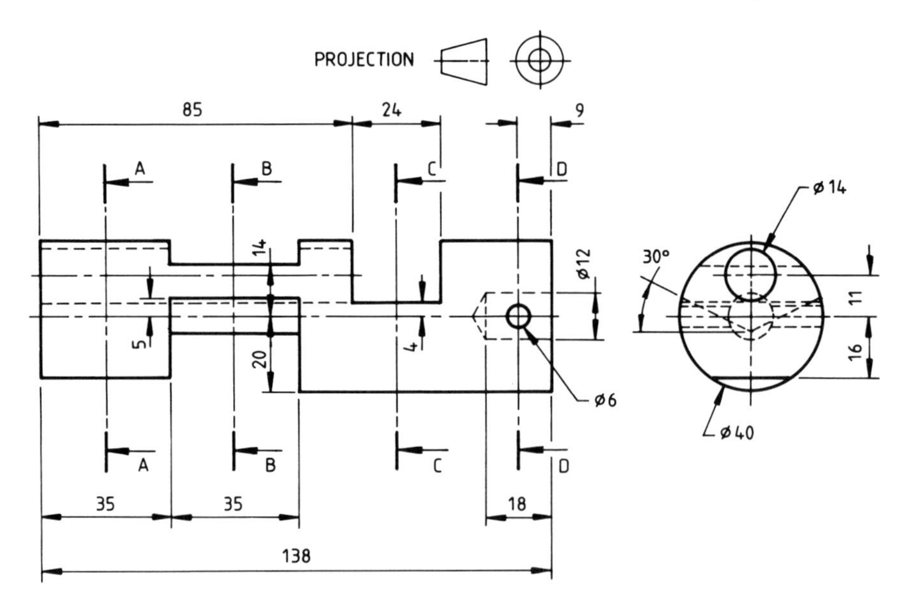

▶**Figure 6.6** Operating valve

Learning Assignment 7

Dimensioning fundamentals

As well as giving a complete shape description of an object using the methods outlined in the previous assignments, a drawing must also give a complete size description; that is, it must be dimensioned. To dimension a drawing correctly and completely requires a knowledge of the standard methods and conventions used to apply dimensions to drawings, and an understanding of the principles governing the selection of the dimensions to be shown. The more commonly used methods and conventions are dealt with in this assignment and Assignment 8, and the selection of dimensions in Assignment 11.

Projection and dimension lines

Projection and dimension lines show the feature to which a dimension refers, and Figure 7.1 illustrates how they are used. These lines are thin and continuous, and for clarity are placed outside the view wherever possible. Projection lines are normally at right angles to the dimension line. They start just clear of the outline of the feature to which they refer, and extend a short way past the dimension line. Dimension lines and projection lines should not cross, but if this is unavoidable then they are generally shown without a break. But if clarity demands it then projection lines (but not dimension lines) may be broken.

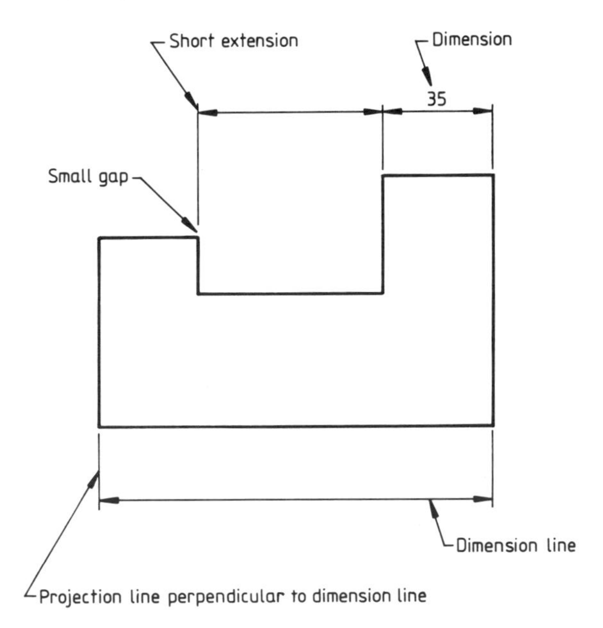

◀**Figure 7.1** Projection and dimension lines

Dimension lines usually end in arrowheads. Arrowheads must be narrow, formed with straight lines and filled in when drawn by hand. Depending on the size of the drawing sheet, they should be between 3 mm and 5 mm long.

POINTS TO NOTE

- Arrowheads drawn by CAD machines need not be filled in, in which case they should be drawn with thin lines.
- The point of the arrowhead must touch the projection line.

Additional information

When tapers are being dimensioned the drawing may be clearer if the projection lines are drawn obliquely. This is illustrated in Part 2 of BS 308.

Examples of the recommended uses of dimension lines are given in Figure 7.2. To make the drawing easier to read, dimension lines are well spaced, and larger dimensions are placed outside smaller dimensions to avoid dimension and projection lines crossing.

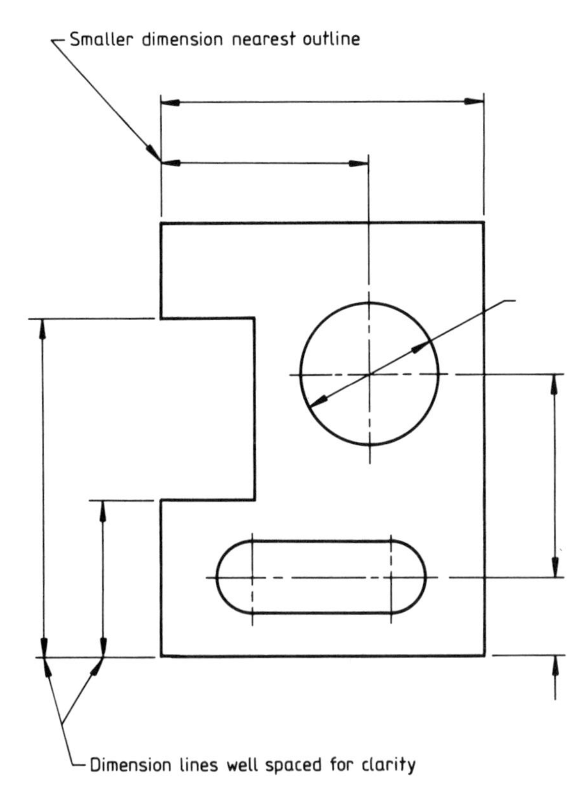

▶**Figure 7.2** Examples of dimension lines

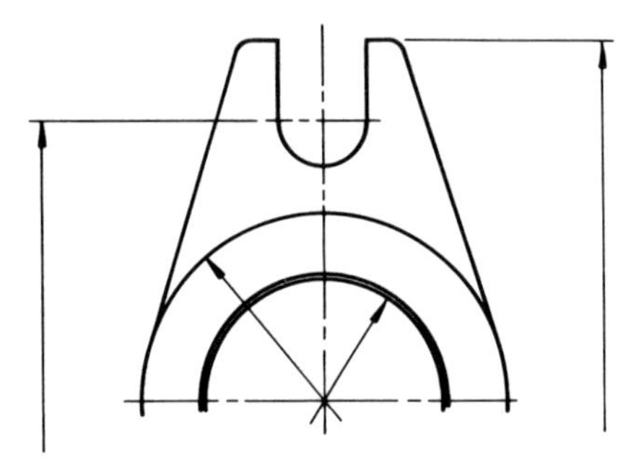

▲**Figure 7.3** Dimension lines on partial view of a symmetrical part

On partial views of symmetrical parts, the portions of the dimension lines should extend a short distance beyond the axis of symmetry, as shown in Figure 7.3. Only one arrowhead is shown, but the full value of the dimension must be given.

Centre lines and their extensions, and continuations of outlines, must never be used as dimension lines. In general, dimension lines should be placed with their arrowheads touching projection lines rather than outlines, but lack of space sometimes prevents this.

Dimensions

Dimensions are given in the form of linear distances, angles or notes.

The dimensions for linear distances are usually in millimetres, for which the unit symbol is 'mm'. To avoid having to show the unit symbol with every dimension, drawings generally carry a note, such as 'Dimensions are in millimetres'.

Linear dimensions must be shown using the least number of significant figures, for example 45 not 45.000. Ways to show toleranced dimensions are given in Assignment 9. Linear dimensions which are less than one millimetre should have a zero before the decimal marker, for example 0.65. The decimal marker is a point which should be bold, have a full letter space and be placed on the base line of the figures. Where there are more than four figures to the left or right of the decimal marker then a full space should separate groups of three figures, counting from the decimal marker, for example 15 750.

Note that in some countries the decimal marker is a comma.

Angular dimensions should be shown in degrees, degrees and minutes, or degrees, minutes and seconds, depending on the accuracy required. Degrees and decimal fractions of a degree may also be used, and are preferred with CAD. Angular dimensions in degrees, minutes and seconds but smaller than one degree should begin with 0°, for example 0° 45' 30".

Place dimensions on the view which shows the features they apply to most clearly. This means, for example, that in general holes are dimensioned and positioned on the view which shows them as circles.

Avoid dimensioning a feature on a view which shows it in hidden detail. If such a view must carry dimensions draw it as a section.

Additional information

During the changeover from imperial to metric measurements which began some years ago, it was common to find drawings with dual dimensioning; that is, the dimensions were given in both inches and millimetres. This practice should now be avoided. However, if conversions must be quoted, the dimensions to which the part is to be made and inspected should be given first, followed by the conversion in parentheses.

Wherever possible, standard material sizes and preferred sizes of items such as screw fasteners, keys, rivets and pins should be used. For example, M12 and M16 hexagon bolts are preferred sizes, and should be selected rather than M14 bolts, which are non-preferred. The use of preferred sizes promotes economy in manufacture, stock holding and distribution because the production of many different sizes of the same article is avoided.

Lettering

It is important that the numerals and letters used for dimension values and notes should be uniform and equally capable of being produced by hand, stencil or machine. The characters must remain legible, not only on a direct photocopy print but also on a reduced-size copy printed back from a microfilm negative or as an image on a microfilm viewing screen. The International Organization for Standardization (ISO) alphabet and figures meet these requirements and have been used throughout this workbook.

Clarity, style, size and spacing are all important, particularly for figures, which unlike letters rarely fall into identifiable patterns and must be read individually.

Character strokes must be black and of consistent density compatible with the line work. Take care that there is enough space between characters and parts of characters to prevent filling in during reproduction of reduced-size copies. Characters must be unambiguous and of open form without embellishments. They may be upright or sloping; the two forms should not be mixed on the same drawing.

A POINT TO NOTE

The smaller angles in sloping letters may make their microfilm reproduction slightly inferior to that from upright letters.

Capital letters are generally used for notes as they are more open and easier to read when reduced in size. BS 308 recommends that lower-case letters should only be used where they are part of a standard symbol, code or abbreviation.

The minimum heights recommended for upper-case characters (i.e. capital letters) are as follows.

Drawing numbers and titles

A0, A1, A2 and A3 drawing sheets – 7 mm.
A4 drawing sheets – 5 mm.

Dimensions and notes

A0 drawing sheets – 3.5 mm.
A1, A2, A3 and A4 drawing sheets – 2.5 mm.

Lower-case characters should have a body height of about 0.6 times the height of upper-case letters.

The thickness of the strokes of the character should be about 0.1 times the height of the character.

The space between lines of lettering should be at least half the character height, but for drawing titles closer spacing is sometimes unavoidable.

Notes

Notes should be lettered parallel to the bottom edge of the drawing sheet, whether this is a long or short edge of the sheet.

Notes of a general character should be grouped together and not spread over the drawing.

A POINT TO NOTE

When preparing a drawing on a CAD machine the dimensioning may be easier to handle if the notes and dimensions are grouped together on their own layers.

Place notes which apply to specific details near to the relevant feature, but not so close that they crowd the view.

Do not underline notes. If a note needs to be emphasized then use larger than normal characters.

Leader lines

Leader lines show where notes and dimensions are intended to apply. They are straight, thin and continuous, and end in either an arrowhead or a bold dot, as shown in Figure 7.4. Leader lines with arrowheads must touch and stop on an outline, as in Figure 7.4(a); those with dots should end within the outline, as in Figure 7.4(b). Horizontal and vertical leader lines should not be used, and leader lines must not pass through the intersection of other lines. For clarity, the slope of leader lines should contrast with that of nearby lines.

Long or intersecting leader lines should be avoided, even if it means repeating dimensions or notes, as in Figure 7.5. Figure 7.6 shows how the use of notes and symbols can avoid long and intersecting leader lines.

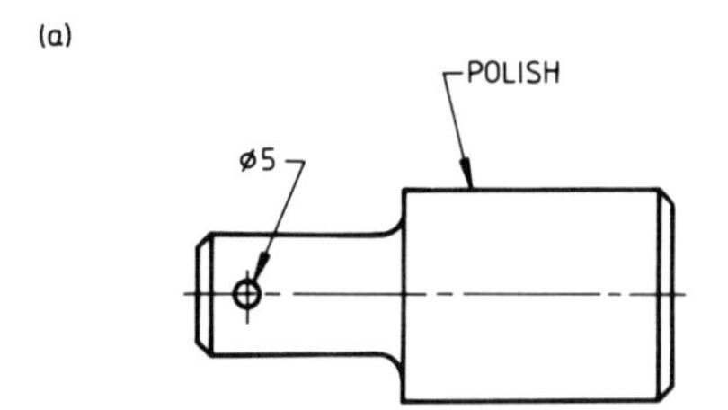

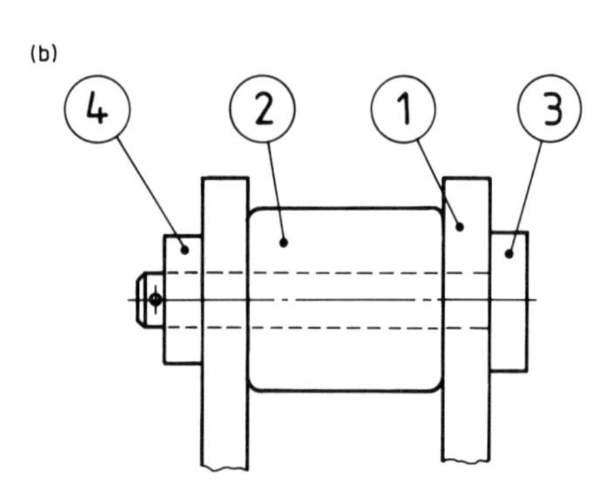

▲**Figure 7.4** Leader lines

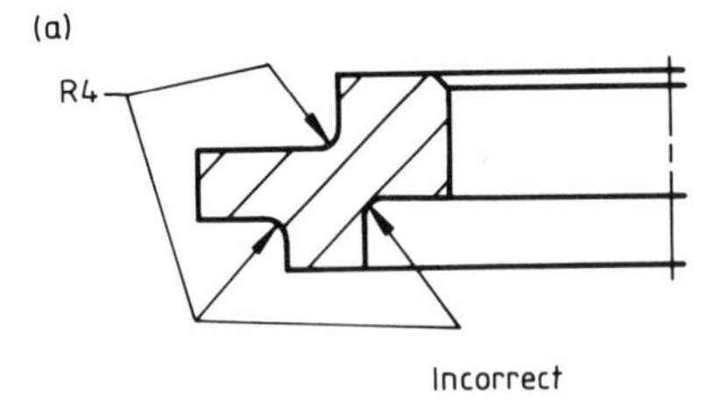

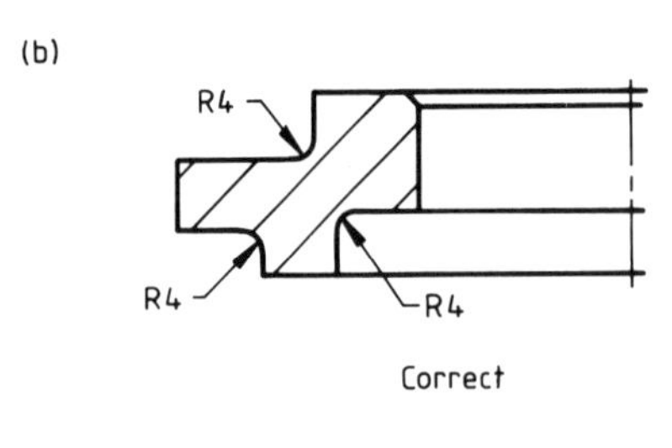

▲**Figure 7.5** Avoiding long and intersecting leader lines

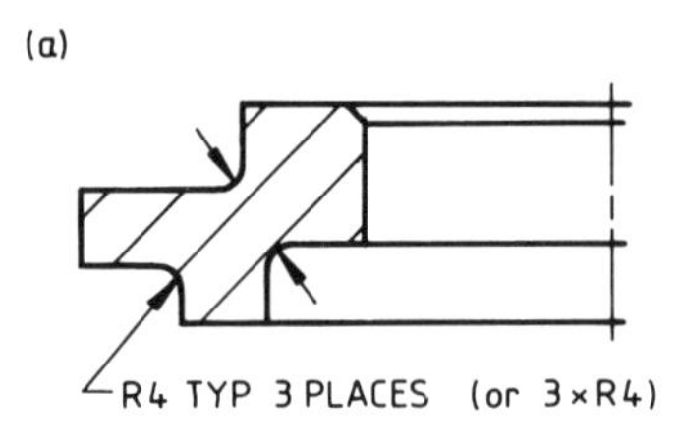

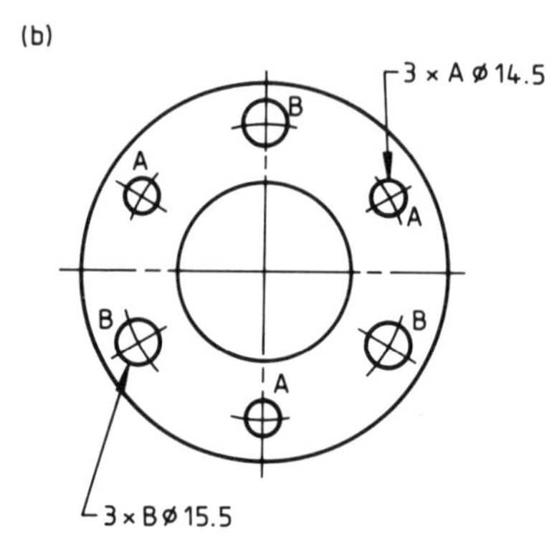

▲**Figure 7.6** Use of notes and symbols to avoid long and intersecting leader lines

Additional information

Figure 7.4(b) also shows 'balloons' containing item references (1, 2, 3, etc.). (See also Figure 10.8.) They are used on assembly drawings to identify the parts which make up the assembly, and are shown in an item or part list. This gives information such as how many of each part is required – the 'number off' – and what its material is to be.

Item references should be at least twice as high as the characters used for dimensions and notes, and be encircled with a thin line. They should preferably be arranged in columns and rows and these should not be placed so close to the view as to crowd it. At the same time, the leader lines should be no longer than is necessary for clarity. The leader lines should preferably end in dots inside the outline of the part, but leader lines with arrowheads touching the outline may have to be used when the parts are small.

The use of item references is covered in detail in Part 1 of BS 308.

Arrangement of dimensions

Dimensions should be positioned so that they can be read from the bottom or right-hand side of the drawing. Figure 7.7 shows how this is done for linear dimensions. Try to keep dimensions outside the hatched zones in the figure. In Figure 7.8(a) the angular dimensions are arranged to be read from the bottom or the right-hand side of the drawing. It is however permissible to position them as in Figure 7.8(b), where they are all readable from the bottom.

Figure 7.8 shows that the dimension line for an angle is a circular arc with its centre on the point of the angle.

Dimensions are generally placed near the middle of the dimension line and slightly above and clear of it, as shown in Figure 7.1, but to avoid them being crossed by other lines they may be moved nearer to one of the arrowheads. Dimensions must not be placed in a gap in the dimension line.

Where there are several parallel diameter dimension lines close together, dimensions may be staggered to avoid congestion, as in Figure 7.9(a) and (b). Sometimes clarity may be improved if the dimension lines are shortened as well, as in Figure 7.9(b).

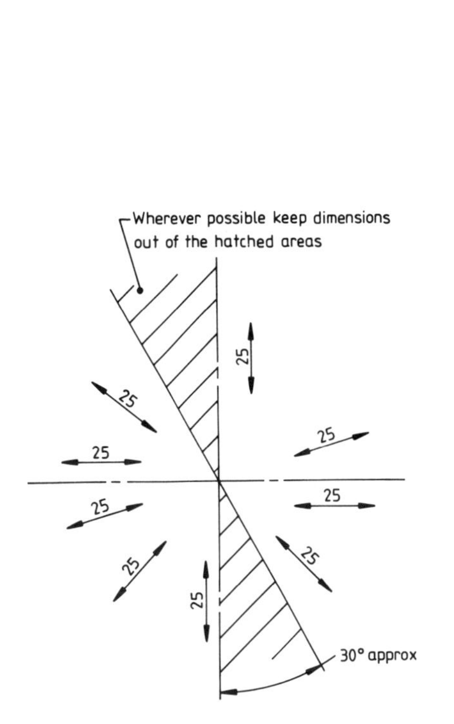

▲Figure 7.7 Orientation of linear dimensions

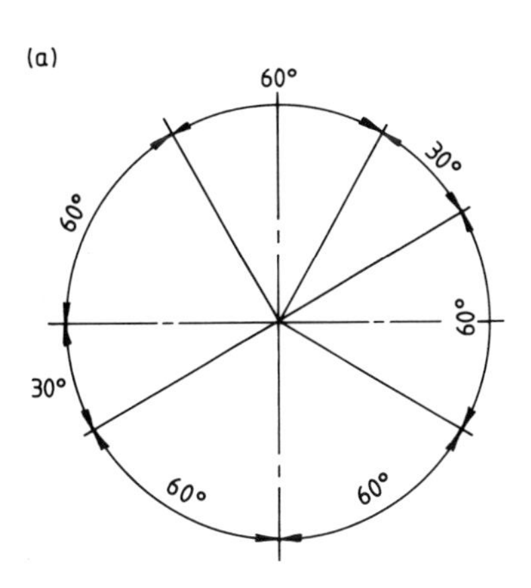

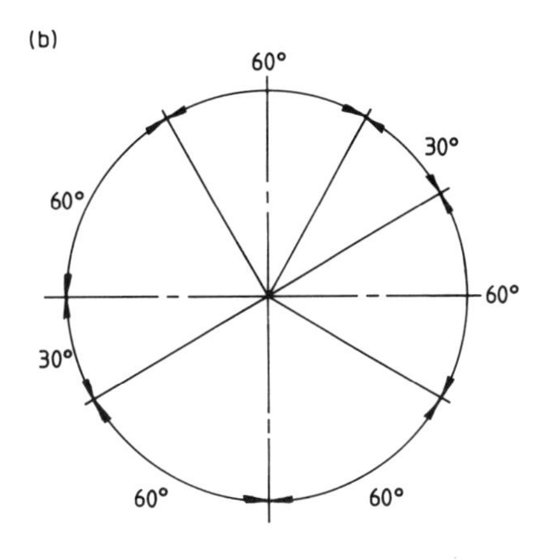

▲Figure 7.8 Orientation of angular dimensions

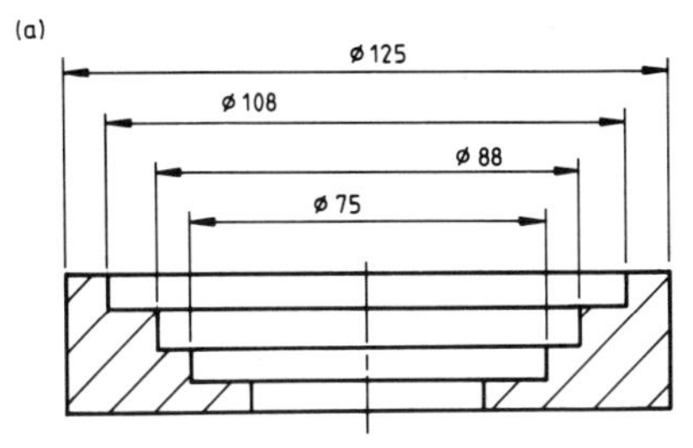

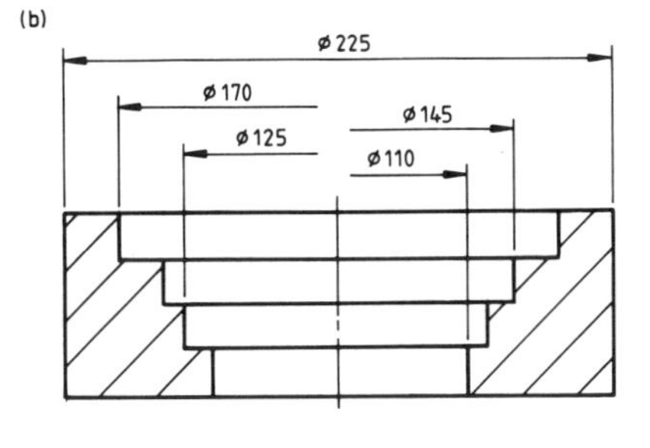

▲Figure 7.9 Arrangement of diameter dimension lines to avoid congestion

Figure 7.10 shows the use of extensions to dimension lines where space is limited.

Where small features have to be dimensioned and there is limited space in which to place the dimension, any of the methods in Figure 7.11 may be used.

Where it is necessary to show that a feature is not drawn to scale, its dimension should be underlined with a thick straight line.

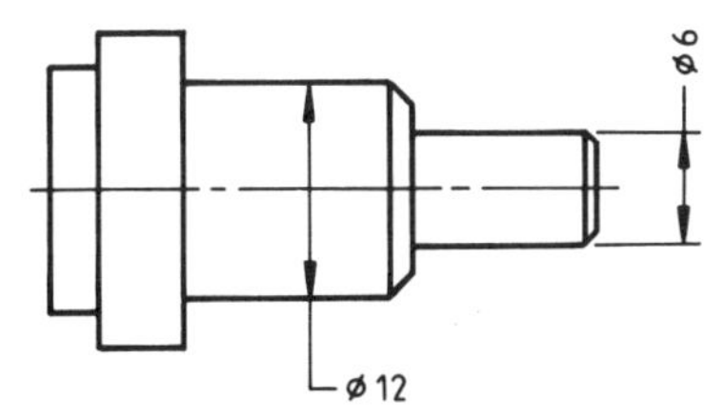

▲Figure 7.10 Use of extended dimension lines

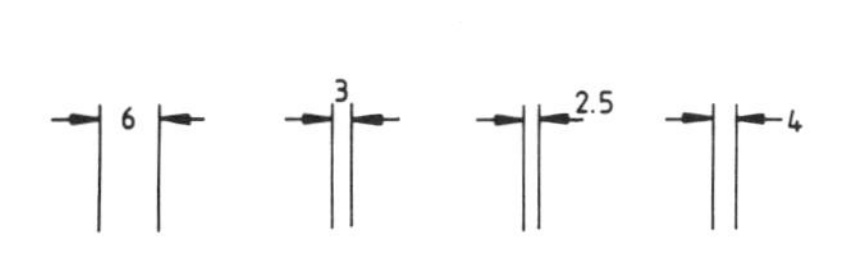

▲Figure 7.11 Dimensioning small features

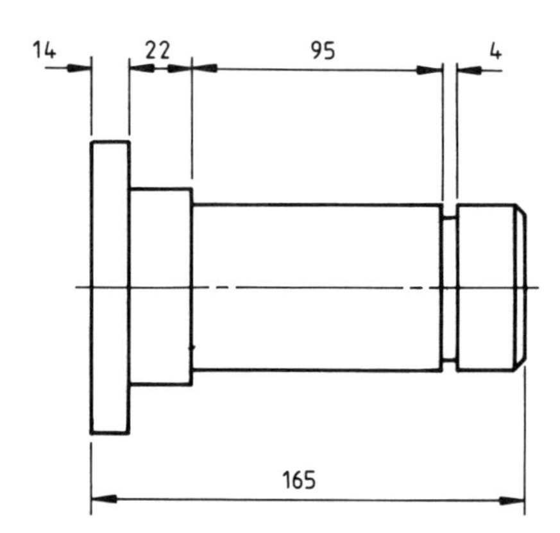

▲**Figure 7.12** Redundant dimension omitted

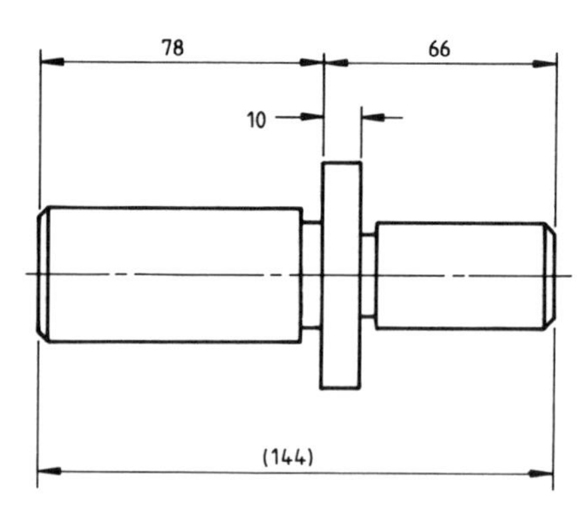

▲**Figure 7.13** Auxiliary dimension

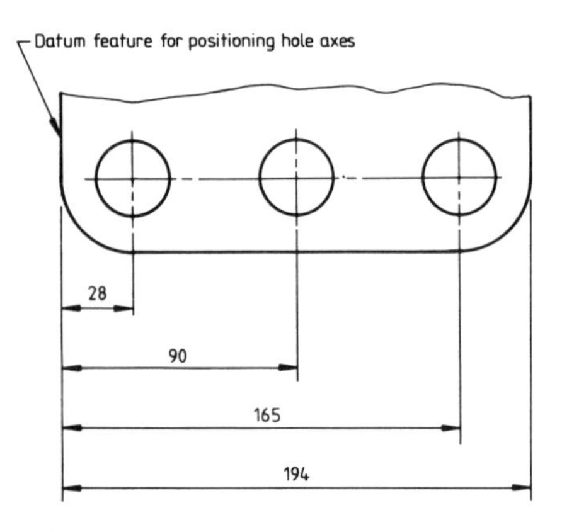

▲**Figure 7.14** Parallel dimensioning

Auxiliary dimensions

In Figure 7.12 the overall length, 165 mm, of the part is given, together with four intermediate length dimensions, 14 mm, 22 mm, 95 mm and 4 mm. The remaining length dimension at the right-hand end of the view is redundant, and therefore is omitted. If this length were given then the overall length would be shown in two different ways, which could lead to confusion.

Sometimes, however, redundant dimensions will provide useful information. Where this is so they are shown as auxiliary dimensions. If all the intermediate dimensions are necessary for the part to function correctly then the overall length may be shown as an auxiliary dimension by enclosing it in parentheses. Figure 7.13 is an illustration. The overall length of a part is sometimes needed to decide how much material to order, or to cut stock into approximate lengths.

Note that an auxiliary dimension is not used in making or inspecting the part, nor does it carry a tolerance. (Tolerances are dealt with in Assignment 9.)

Dimensioning methods

Dimensions can be arranged parallel to each other, or in a chain, or a combination of both these methods can be used. Sometimes a coordinate method may be appropriate.

Parallel dimensioning

Parallel dimensioning is illustrated in Figure 7.14. A number of dimensions originate from a datum feature which is a face, an edge or an axis. In Figure 7.15 the left-hand edge of the part is the datum feature for positioning the three hole axes.

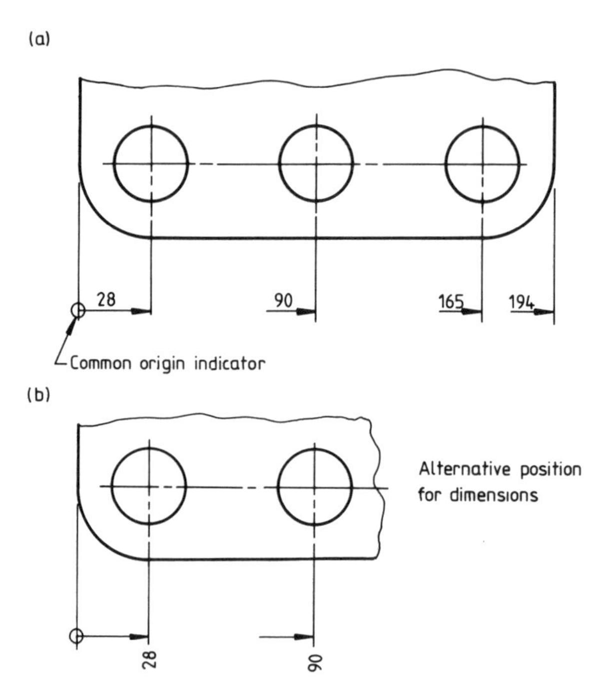

▶**Figure 7.15** Superimposed running dimensioning

Superimposed running dimensioning

This is a simplified form of parallel dimensioning and is shown in Figure 7.15. It is useful when there is not enough space for all the parallel dimension lines which are necessary. The datum feature is identified by the common origin indicator, which is a circle of about 3 mm in diameter drawn with a thin line. Figure 7.15(a) shows the usual positioning of the dimensions, but the positioning in Figure 7.15(b) is useful where there are only small differences in the dimensions, leaving insufficient space to insert the values.

Chain dimensioning

Chain dimensioning is illustrated in Figure 7.16. As explained later in Assignment 9, this method should only be used where the possible accumulation of tolerances does not endanger the correct functioning of the part.

▲**Figure 7.16** Chain dimensioning

Combined parallel and chain dimensioning

Figure 7.17 is an example of combined parallel and chain dimensioning, and is the dimensioning method most commonly used. The left-hand edge of the part is the datum feature used to position the axes of the larger holes. These in turn are the datum features for the axes of the smaller holes.

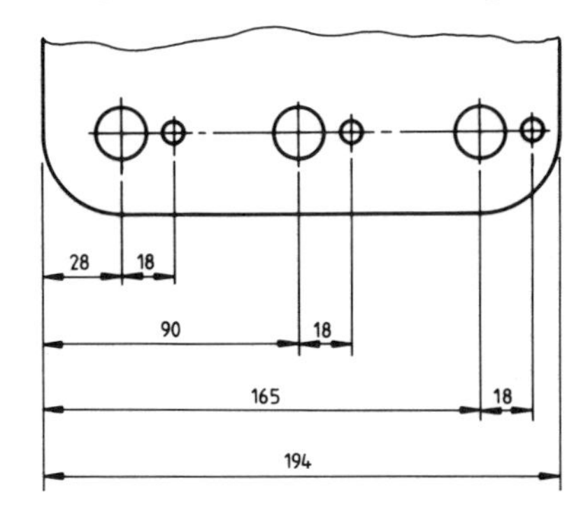

▲**Figure 7.17** Combined parallel and chain dimensioning

Dimensioning by coordinates

For some parts, such as the plate in Figure 7.18, much space may be saved if dimensioning by coordinates is used. This method uses superimposed running dimensioning in two directions at right angles. The common origin can be any suitable datum feature.

Figure 7.19 shows that the coordinate dimensions may be tabulated instead of applying them as in Figure 7.18. Where this can be done the absence of dimension and projection lines will simplify the drawing. However, where several groups of features such as holes are required on the part, it will generally be necessary to identify each group separately, as has been done here.

Symbols and abbreviations

Symbols and abbreviations are used on drawings to save space and time. Provided that those used are commonly understood, the dimensioning will

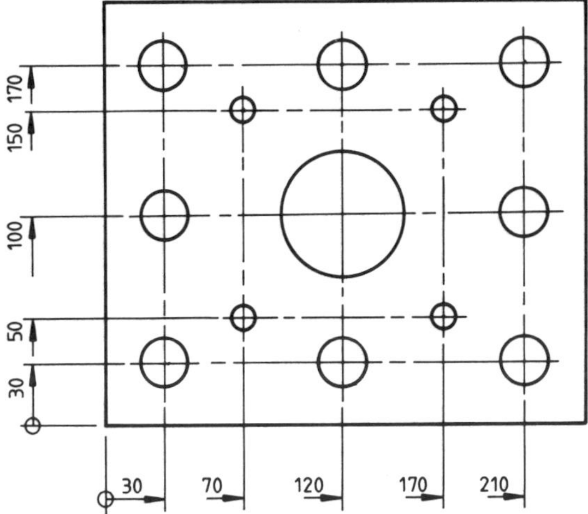

▲**Figure 7.18** Dimensioning by coodinates

▼**Figure 7.19** Dimensioning by coodinates and a table

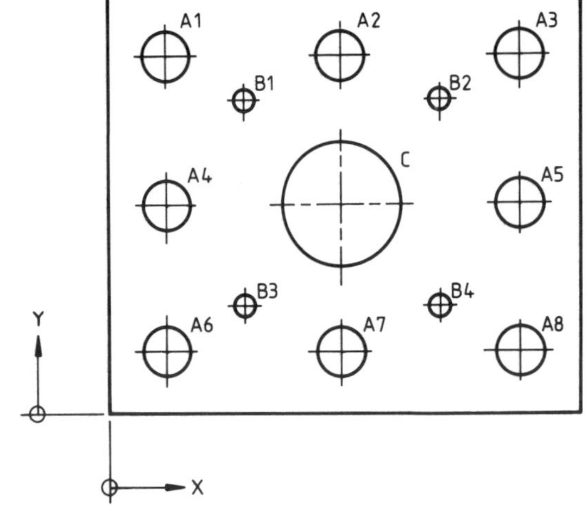

HOLE	X	Y	∅
A1	30	170	25
A2	120	170	25
A3	210	170	25
A4	30	100	25
A5	210	100	25
A6	30	30	25
A7	120	30	25
A8	210	30	25
B1	70	150	12
B2	170	150	12
B3	70	50	12
B4	170	50	12
C	120	100	60

POINTS TO NOTE

- Abbreviations are the same in the singular and plural, thus the shortened form DRG is used for 'drawing' and 'drawings'.
- Abbreviations do not end with a full stop, except where they make a word themselves, as with 'NO.' for number.

be simplified as well, without detriment to precise and clear description. Symbols and abbreviations which are not in common use should be avoided, and the intended meaning stated in words.

The following commonly accepted symbols and abbreviations have been selected from the list in BS 308 : Part 1.

Term	Symbol or abbreviation
Across flats	AF
Assembly	ASSY
Centres	CRS
Centre line	
on a view and across the centre line	℄
in a note	CL
Centre of gravity	CG
Chamfered or chamfer *(in a note)*	CHAM
Cheese head	CH HD
Countersunk	CSK
Countersunk head	CSK HD
Counterbore	CBORE
Cylinder or cylindrical	CYL
Diameter	
in a note	DIA
preceding a dimension	∅
Dimension	DIM.
Drawing	DRG
Equally spaced	EQUI SP
External	EXT
Figure	FIG.
Hexagon	HEX
Hexagon head	HEX HD
Insulated or insulation	INSUL
Internal	INT
Left hand	LH
Long	LG
Machine	MC
Material	MATL
Maximum	MAX
Minimum	MIN
Not to scale *(in a note)*	NTS
Number	NO.
Pitch circle diameter	PCD
Radius	
in a note	RAD
preceding a dimension	R
Reference	REF
Required	REQD
Right hand	RH
Screw or screwed	SCR
Sheet *(referring to a drawing sheet)*	SH
Specification	SPEC
Spherical diameter *(only preceding a dimension)*	S∅
Spherical radius *(only preceding a dimension)*	SR
Spotface	SFACE
Square	
in a note	SQ
preceding a dimension	□ or ⊠
Standard	STD
Taper *(on diameter or width (see Figure 8.15(b))*	[taper symbol]
Thread	THD
Thick	THK
Tolerance	TOL
Typical or typically	TYP
Undercut	UCUT
Volume	VOL
Weight	WT

Selected surface texture symbols are given in Assignment 15.

A POINT TO NOTE

The symbol for 'square' without the diagonals is recognised internationally through ISO. The only national standard to recommend the symbol with the diagonals is BS 308.

Additional information

Symbols and abbreviations for units of measurement and physical quantities are defined in the appropriate British Standards and should be used on drawings where necessary.

Further recognised abbreviations are listed in other British Standards dealing with particular subjects.

There are many British Standards which set out symbols used by a particular engineering discipline or in a particular industry. Such symbols should be used with care, and only after consulting the relevant standard. Examples of such standards are:

- BS 499 Welding Terms and Symbols.
- BS 3939 Graphical Symbols for Electrical Power, Telecommunications and Electronic Diagrams.
- BS 2917 Graphical Symbols Used on Diagrams for Fluid Power Systems and Components.

TASKS

Task 7.1

Redraw the side plate in Figure 7.20 and correct all the dimensioning errors.

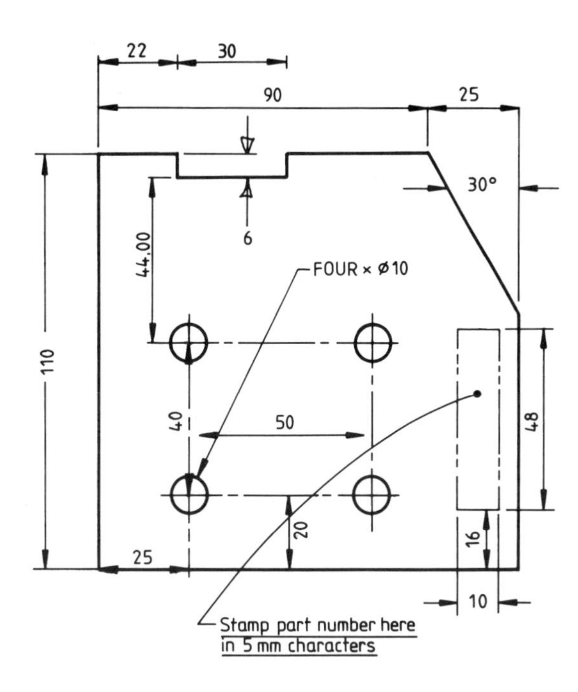

◀**Figure 7.20** Side plate

Task 7.2

Draw sufficient fully dimensioned orthographic views of the bracket in Figure 7.21 to describe it completely.

Task 7.3

Provide a fully dimensioned drawing of the 3 mm thick locking strip in Figure 7.22 to satisfy the following requirements.

(a) The centre lines of the slots A to E and tabs F to J are to be positioned using superimposed running dimensioning from edge K as the common origin.

(b) The distance from edge K to slot A is 25 mm; between slots A and B, and D and E it is 20 mm; and between slots B and C, and C and D it is 30 mm. From edge K to edge L is 10 mm, and from edge K to edge M is 150 mm.

(c) The slots are 20 mm deep and 5 mm wide.

(d) From edge K to the centre line of tab F is 35 mm; the distances

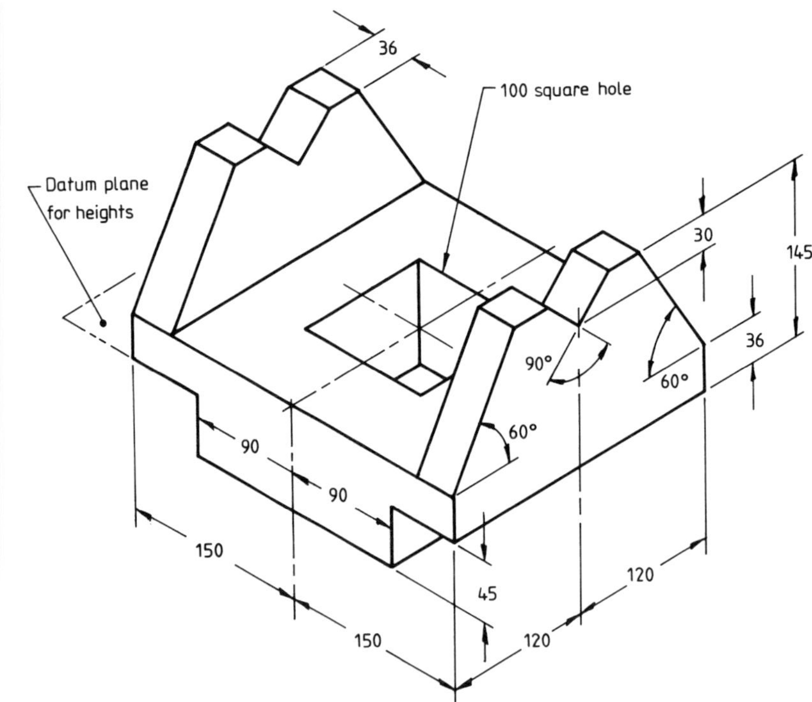

▶**Figure** 7.21 Bracket

between tabs F and G, and H and J, are 28 mm; and that between G and H is 24 mm.

(e) The tabs are 10 mm high and 10 mm wide.

(f) The datum plane for heights is X. From plane X to planes Y and Z is 30 mm and 18 mm, respectively.

Complete the drawing by adding the following notes:

1. Break all sharp corners 0.2 min 0.4 max.
2. Cadmium plate 0.04 thk to Spec 019/92.

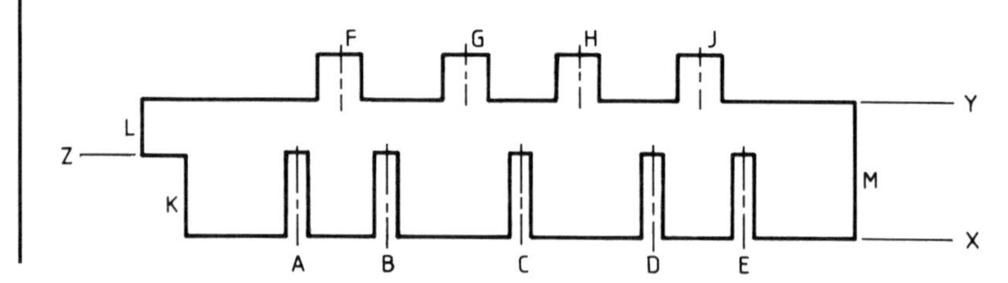

▶**Figure** 7.22 Locking strip

Learning Assignment 8

Methods of dimensioning common features

This assignment deals only with those features which are met most often. BS 308 : Part 2 should be consulted for a full treatment of the topic.

Circles and diameters

Complete circles must always be dimensioned by their diameters, not their radii, using one of the methods in Figure 8.1. The symbol Ø, meaning diameter, precedes the dimension. Circles must be shown with two centre lines.

In Figure 8.1 (d) do not add an arrowhead touching the circle to the extension of the dimension line. Do not replace the extension by a leader line. These points apply equally to radii dimensions such as that in the left-hand example in Figure 8.4(a).

Figure 8.2 shows several ways for dimensioning diameters. The dimension should be placed on the view which provides the maximum clarity, as in Figure 8.2(a). Here the diameter dimensions are placed on the lengthwise view because the other has several closely spaced circles. In Figure 8.2(b) the view is simplified not by using the normal projection and

▼**Figure 8.2** Dimensioning diameters

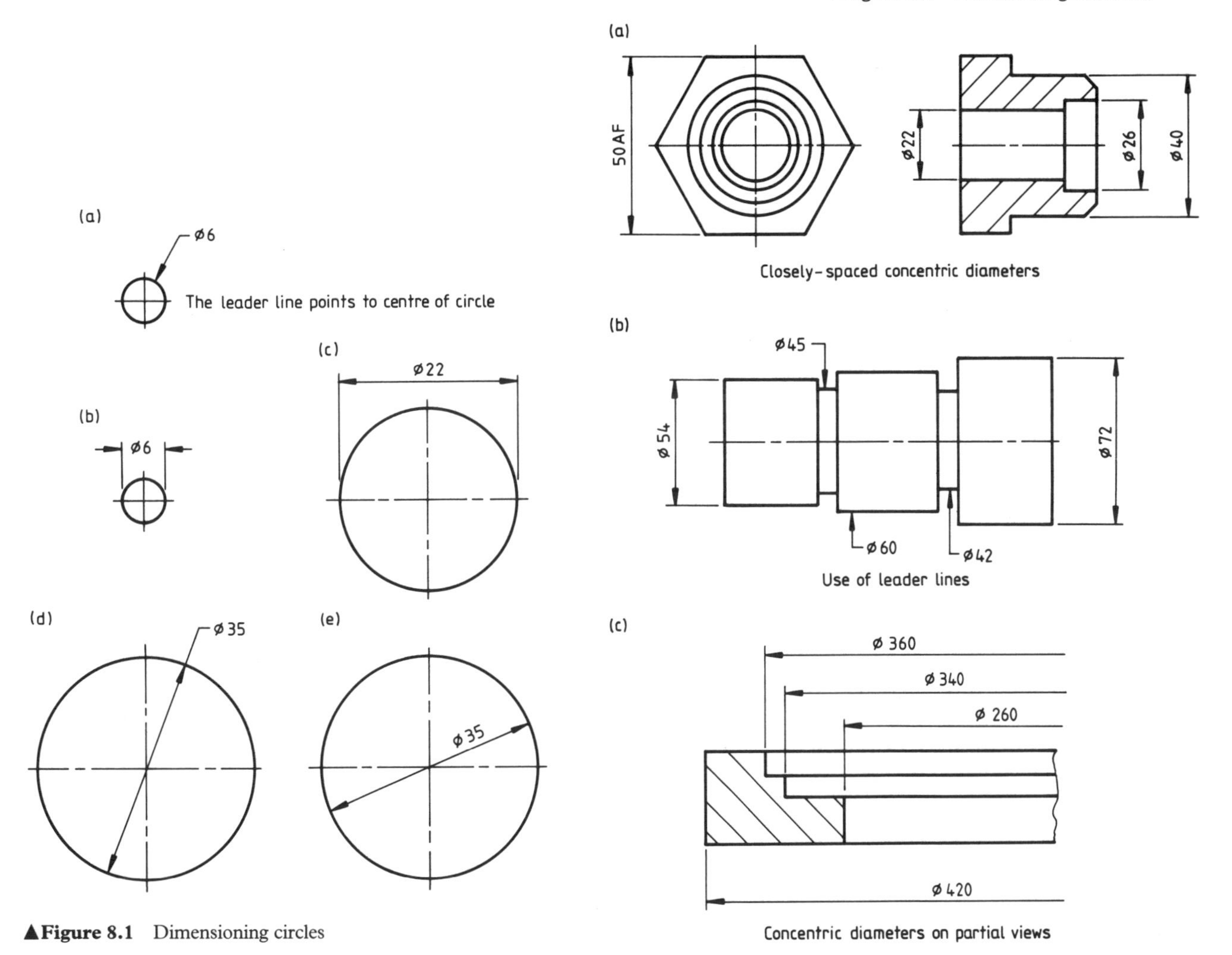

▲**Figure 8.1** Dimensioning circles

dimension lines but rather, by relating the diameter dimensions to the features by leader lines. Figure 8.2(c) shows how staggered diameter dimensions may be used on partial views.

The diameter of a spherical surface should be dimensioned as shown in Figure 8.3.

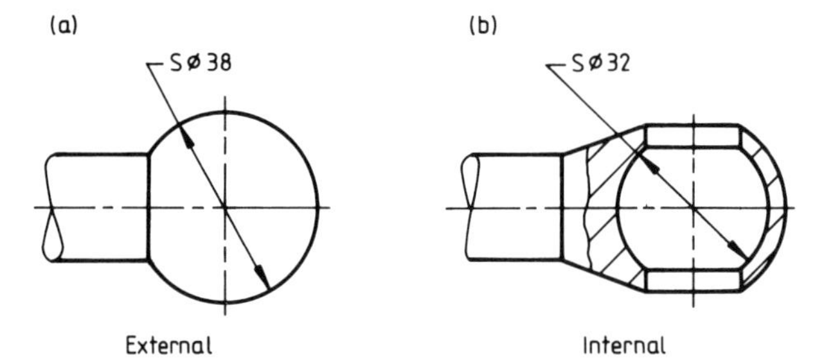

▶**Figure 8.3** Dimensioning spherical diameters

Radii

The dimension line for a radius passes through, or is in line with, the centre of the arc. The dimension line carries one arrowhead only, and this touches the arc. The symbol R, meaning radius, precedes the dimension. These points are illustrated in Figure 8.4. Figure 8.4(a) shows the dimensioning of arcs whose centres are located, and Figure 8.4(b) the dimensioning when the arc centre does not need to be located.

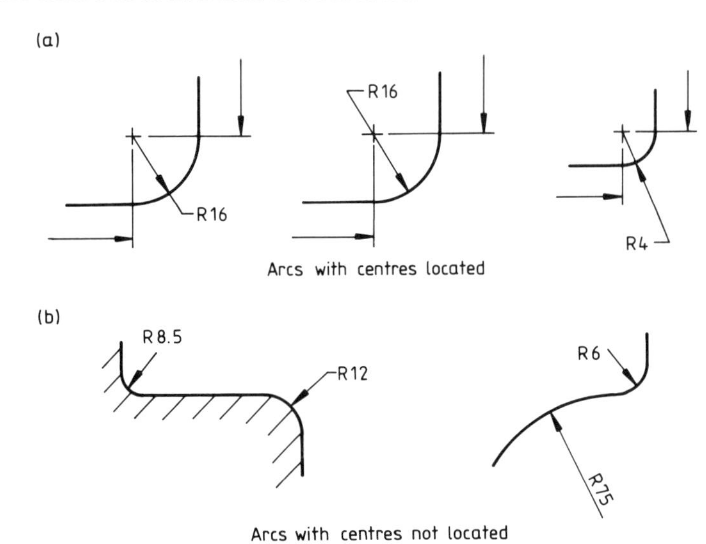

▶**Figure 8.4** Dimensioning radii

The fillet radii and rounds on external corners of castings and forgings should be dimensioned by a general note rather than individually. Examples are 'Fillets and rounds R4' and 'Unspecified radii to be R5'.

If the size of a radius is fixed by another dimension, as in Figure 8.5, it is indicated by the symbol R without a dimension. In the given example the width of the feature is 60 mm, so the radius can only be 30 mm.

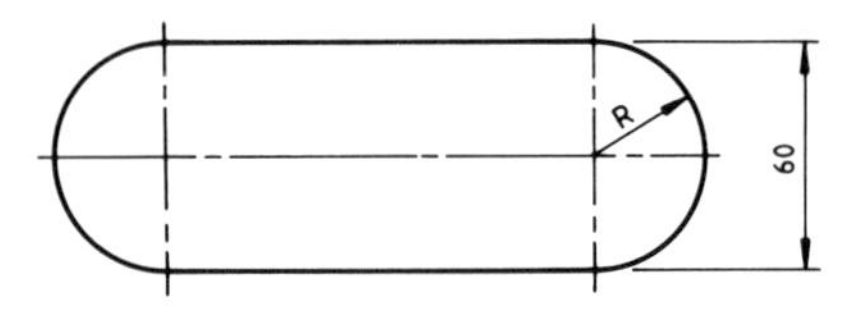

▲**Figure 8.5** Radius established by another dimension

Figure 8.6 shows how spherical radii are dimensioned.

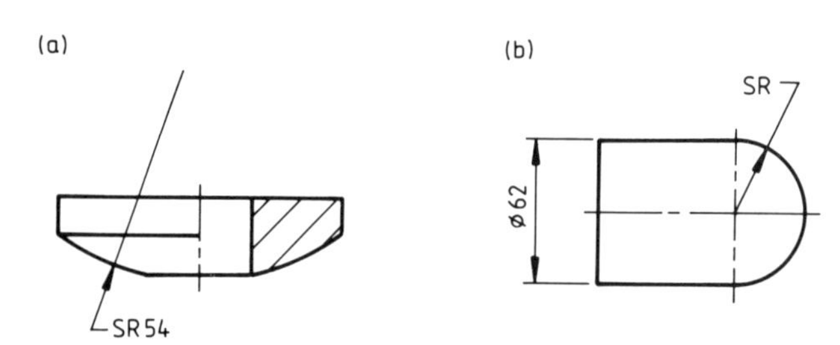

▶**Figure 8.6** Dimensioning spherical radii

Size of holes

Typical methods of dimensioning hole sizes are shown in Figure 8.7. The production method, such as drill, ream, punch, core, etc., should be specified only if a particular method is essential for the part to function satisfactorily. The use of the word 'hole' in the dimensioning is unnecessary.

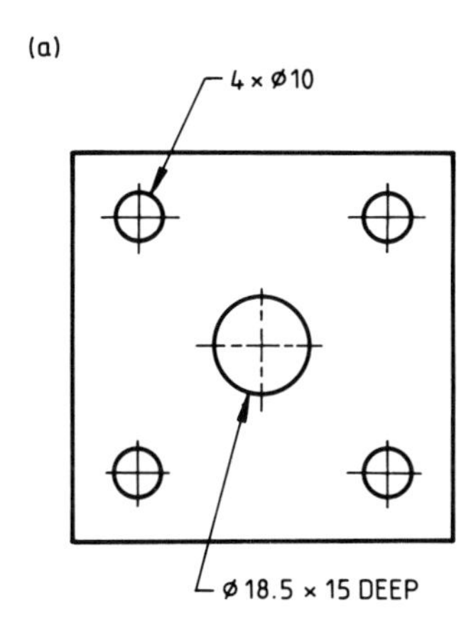

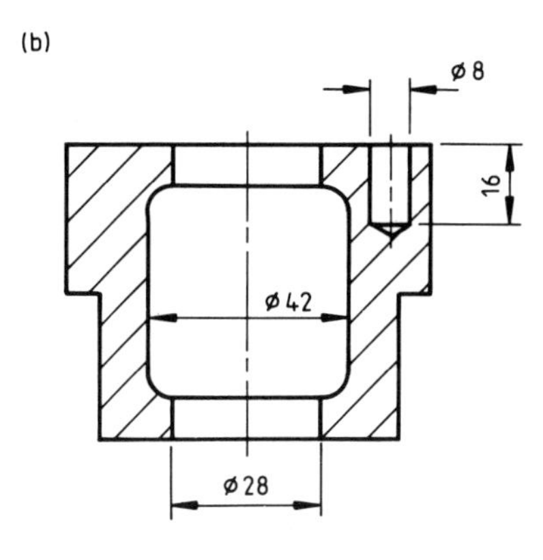

◄**Figure 8.7** Dimensioning holes

POINTS TO NOTE

- When the depth of a hole is given in a note, it refers to the depth of the cylindrical portion of the hole, not the depth to the point of the drill. See Figure 8.7(a).
- When no drill depth is specified, as with the four 10 mm diameter holes in Figure 8.7(a), the holes are required to go right through the part.

Positioning of holes

Figure 8.8 shows how hole positions should be dimensioned when they are located by angular spacing around a circle. When the holes are positioned by centre distances or coordinates the dimensioning in Figure 8.9 should be followed. If the holes are positioned by coordinates, although in fact they lie on an arc or a circle as in Figure 8.9, the pitch circle radius or pitch circle diameter should be given as an auxiliary dimension.

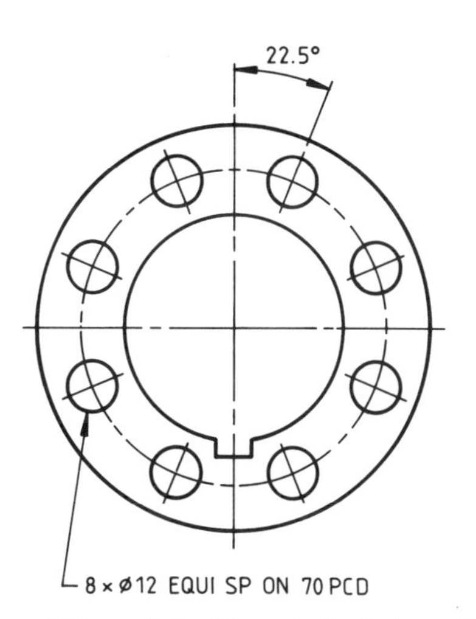

▲**Figure 8.8** Dimensioning hole positions with equal angular spacing

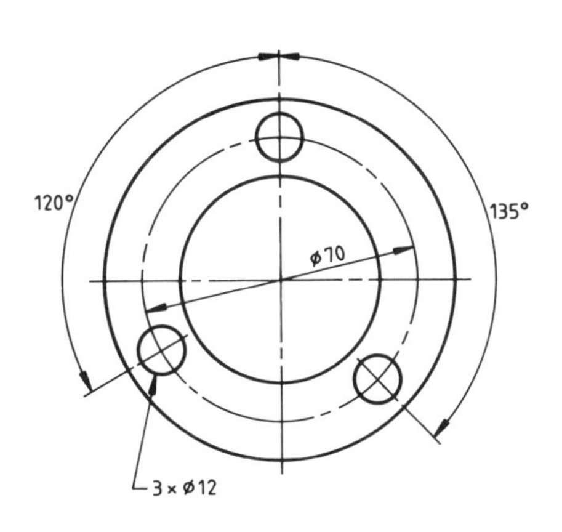

▲**Figure 8.9** Dimensioning hole positions with unequal angular spacing

Chamfers

45° chamfers are the most common size, and ways of dimensioning them are given in Figure 8.10(a), (b) and (c). Chamfers at angles other than 45° should be dimensioned as in Figure 8.10(d), (e) and (f).

A POINT TO NOTE

To avoid any possibility of confusion between the dimension along the chamfer and the axial dimension, 45° chamfers should not be dimensioned by a note and leader line.

▶**Figure 8.10** Dimensioning chamfers

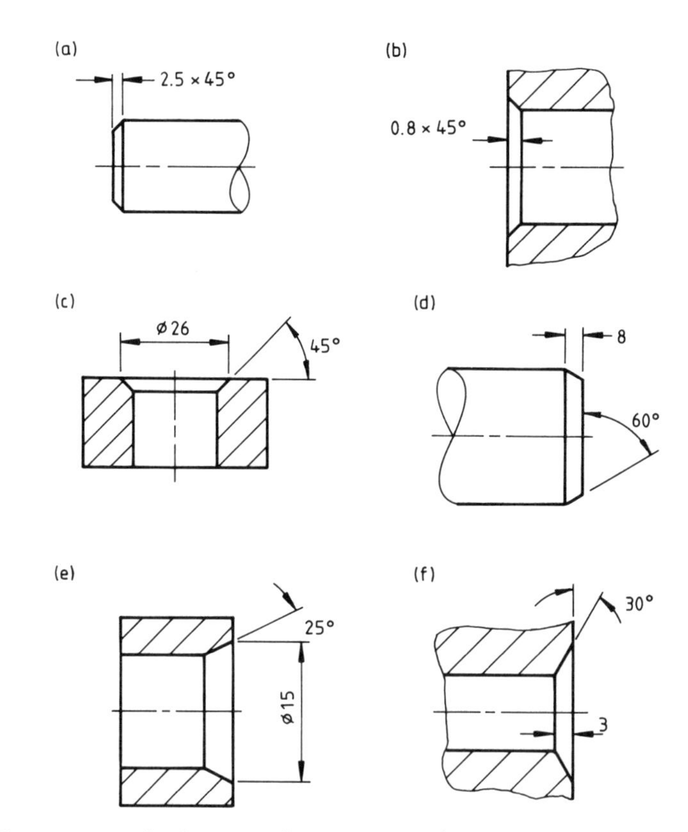

> *POINTS TO NOTE*
> - When countersinks and counterbores are dimensioned by a note which includes the diameter of the hole, as in Figure 8.11(a) and Figure 8.12(a), the arrowhead on the leader line must touch the surface of the hole. This also applies to spotfaces; see Figure 8.13(a).
> - The depth of a counterbore is given by the dimension which follows the diameter.
> - Do not use notes such as 'Counterbore to suit M6 ch hd screw'. They are imprecise and cause confusion.

> *A POINT TO NOTE*
> Do not give a depth dimension for a spotface. The term implies that the depth required is the minimum necessary to provide a completely machined surface of the specified diameter.

Countersinks and counterbores

Methods of dimensioning countersinks are illustrated in Figure 8.11. Counterbores should be dimensioned by one of the methods in Figure 8.12.

Spotfaces

Examples of the dimensioning of spotfaces are shown in Figure 8.13.

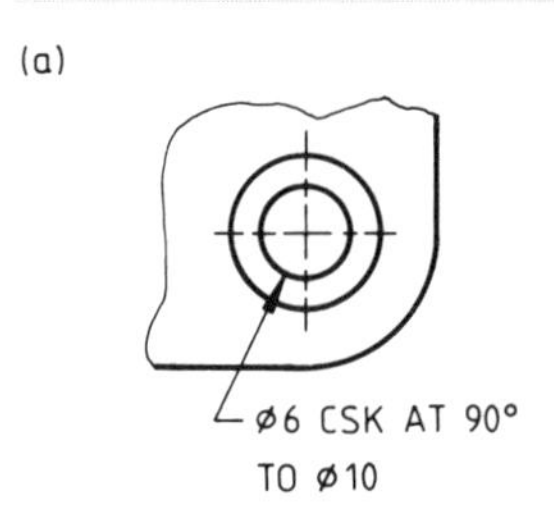

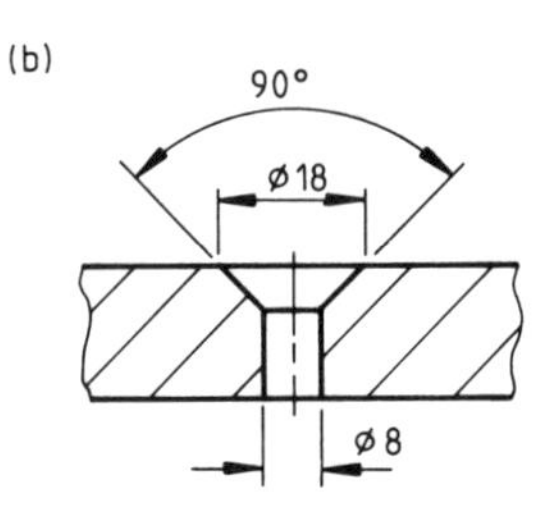

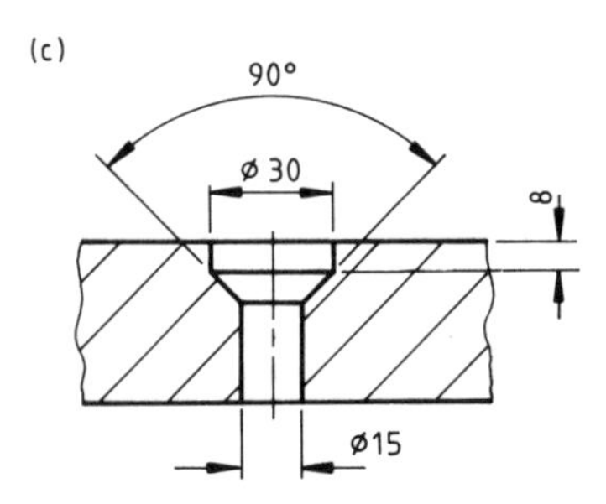

▲**Figure 8.11** Dimensioning countersinks

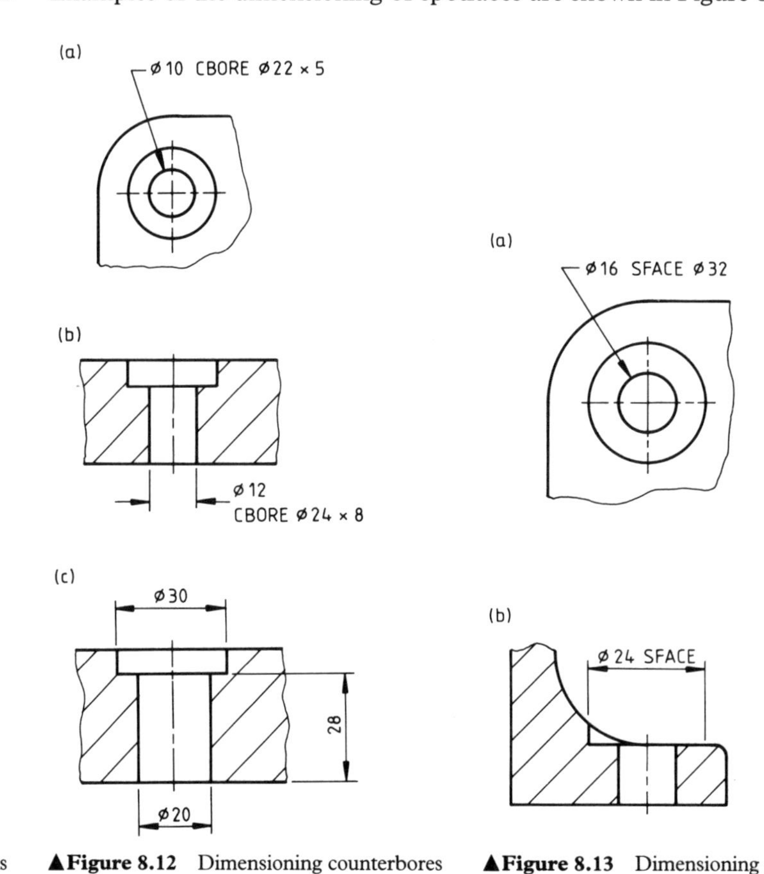

▲**Figure 8.12** Dimensioning counterbores

▲**Figure 8.13** Dimensioning spotfaces

Keyways

Methods for dimensioning keyways in shafts and hubs are given in Figure 8.14. Note that the depth dimensions for keyways other than those for Woodruff keys are given from the far side of the shaft or hub. These are much easier to check than those from the centre line.

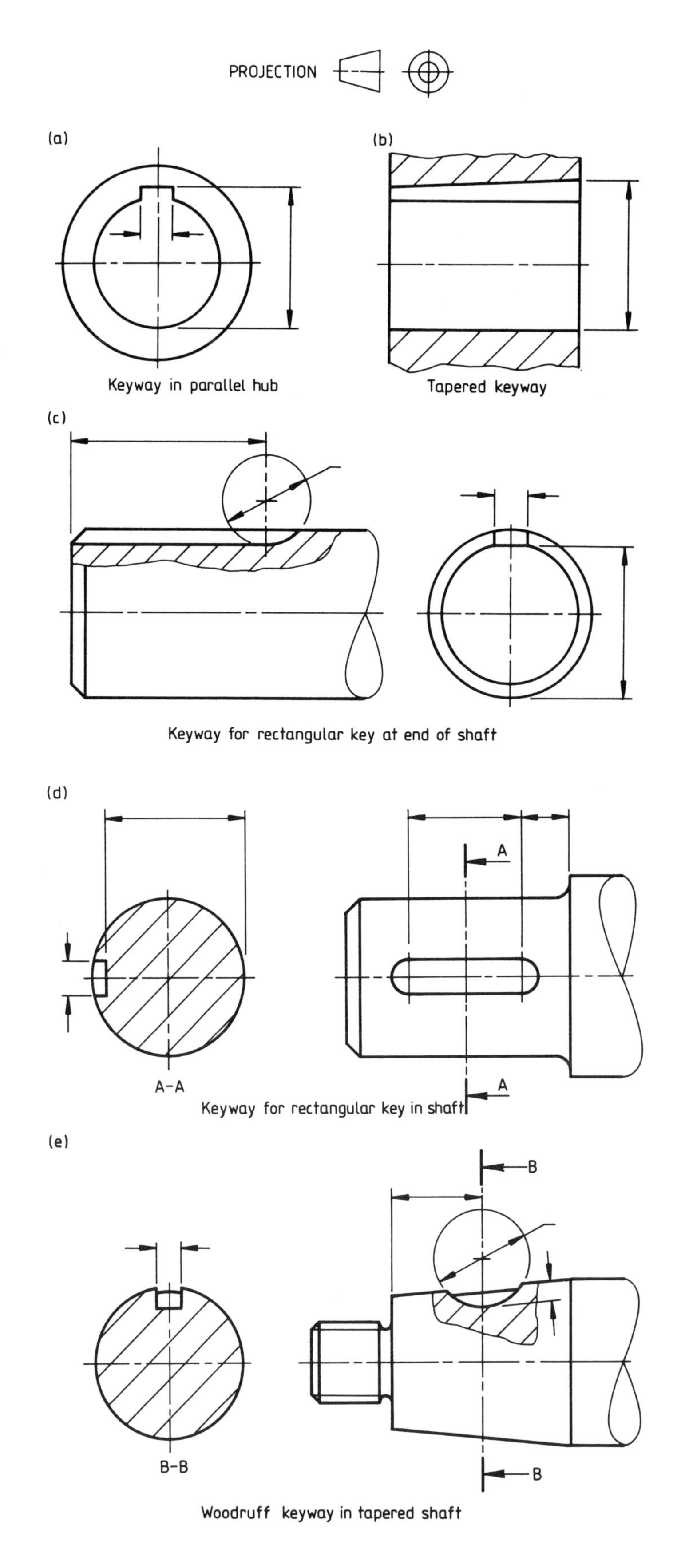

Figure 8.14 Dimensioning keyways

Additional information

Keys are used to prevent relative rotation between shafts and wheels, gears, couplings, cranks and similar parts which are attached to the shafts. They are of steel and fitted partly in the shaft and partly in the hub. Some keys, called feathers, allow axial movement. They are found in gearboxes, for example, and enable the gears to slide along the shafts into and out of mesh. BS 4235 gives information on the types of metric keys and keyways.

For heavy duty, keys are replaced by splines (grooves) cut in both shaft and hub and arranged to fit one inside the other. BS 5686 covers straight-sided splines and BS 6186 involute splines.

Tapered features

Generated tapers

These tapers are generated round an axis and all cross-sections at right angles to the axis are circles.

The size and form of generated tapers are indicated by suitable combinations of the following dimensions:

(a) the diameters at the ends of the tapered feature;
(b) the length of the tapered feature;
(c) the position and diameter at a selected cross-sectional plane at right angles to the axis, which is usually inside the tapered feature, but may be outside;
(d) the rate of taper or the included angle.

The rate of taper is the ratio of the difference in the diameters of two circular sections of a cone to the distance between them.

All the dimensions are illustrated in Figure 8.15(a). No more of them should be given than are necessary to specify the taper completely.

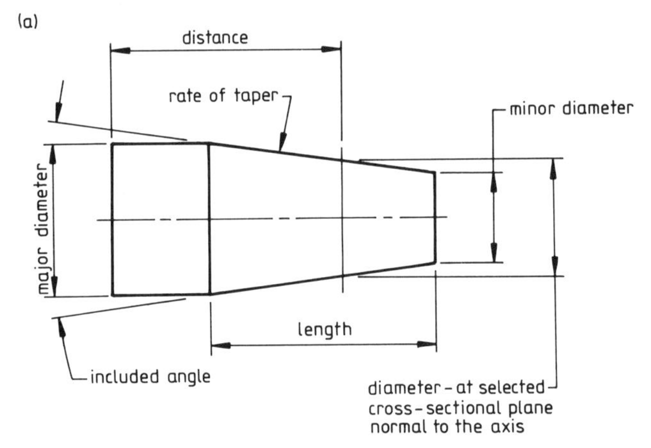

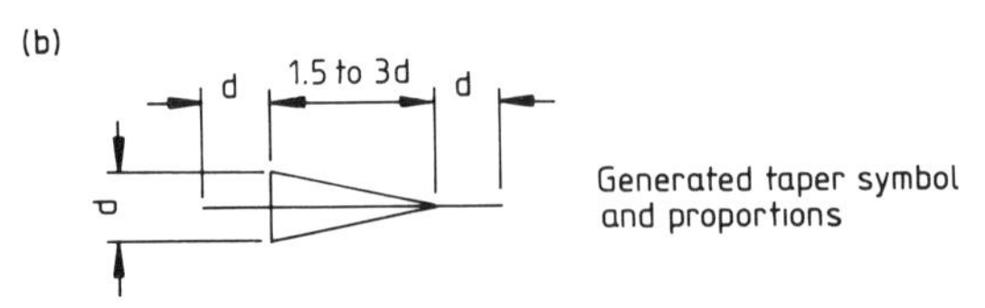

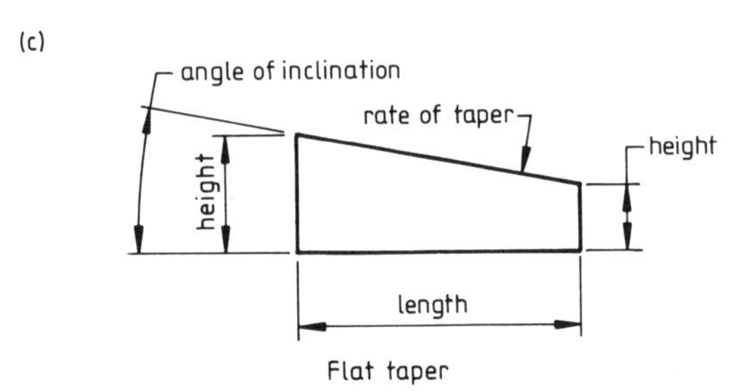

▶**Figure 8.15** Dimensions defining tapered features

Taper symbol
Figure 8.15(b) gives the form and proportions for a symbol which indicates a generated taper. This symbol is recognised internationally and replaces the word 'taper' in such notes as 'Taper 1:6'. The axis (or reference line) of the symbol is to be parallel to the axis of the taper, and the symbol is to point in the direction of the taper on the feature.

Flat tapers

Tapered features of this type consist of an inclined plane related to some base or datum surface, but without any relationship to an axis.

The size and inclination of flat tapers are indicated by suitable combinations of the following dimensions:

(a) the heights at the ends of the tapered feature;
(b) the length of the tapered feature;
(c) the position and height of a cross-sectional plane which is usually within the tapered feature, but may be outside;
(d) the rate of taper or angle of inclination to the base or datum surface.

The rate of taper is the ratio of the difference in the heights of two sections of a flat taper to the distance between them.

Figure 8.15(c) illustrates all the dimensions. No more than the minimum necessary to specify the taper completely should be given.

Additional information

The foregoing notes set out the basic principles only for dimensioning tapered features. The selection of dimensions to specify the required accuracy in different circumstances is outside the scope of this workbook, but is discussed in detail in BS 308 : Part 2.

TASKS

Task 8.1

Draw the given views of the part in Figure 8.16, changing the elevation to a full section and adding all dimensions. The views are to include the following features:

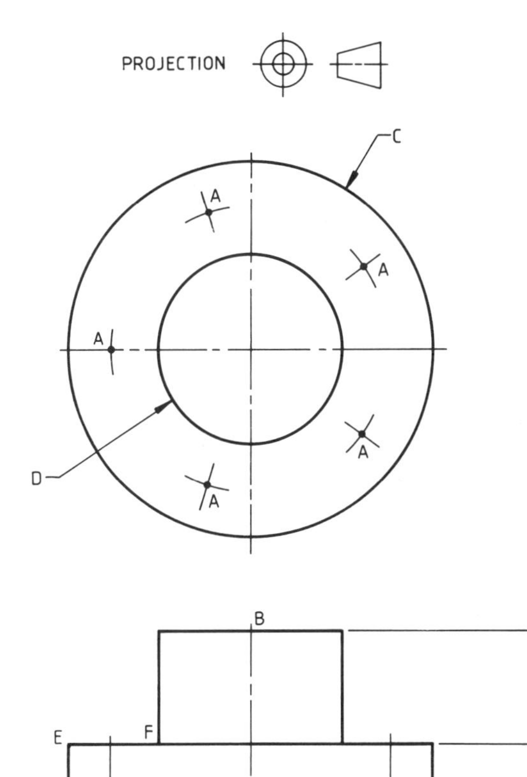

◄**Figure 8.16** Cover

(a) At the five points A, through holes 18 mm in diameter are drilled, equally spaced on a pitch circle of 168 mm diameter. The holes are to be countersunk at 90° to 30 mm in diameter on plane Y.
(b) On axis B a through hole 45 mm in diameter is bored and counterbored at each end to 75 mm diameter for a depth of 18 mm.
(c) 8 mm radii are formed on corners E and F.

Diameters C and D are 220 mm and 110 mm, respectively. Planes Y and Z are 44 mm and 112 mm, respectively, above plane X, which is the datum feature from which heights are to be dimensioned.

Task 8.2

Draw suitable fully dimensioned views of the spacer in Figure 8.17, adding the following omitted features:

(a) At the four points A show through holes drilled 12 mm diameter and spotfaced 20 mm diameter on plane Y.
(b) At the two points D show holes drilled 14 mm diameter for a depth of 40 mm and counterdrilled 22 mm diameter for a depth of 16 mm.
(c) On axis C show a through hole drilled and reamed 30 mm diameter, and countersunk at 90° to 38 mm diameter at plane Z.
(d) At the two points E show holes drilled through at 10 mm diameter with a 3 mm chamfer at 45° on plane X.

Points A are centres for arcs of radius 14 mm forming part of the outer profile. Arcs centred at points B blend with these arcs.

16 mm radii with centres at points D form the ends of the inner profile. The top and bottom of the profile consists of parts of a 64 mm diameter circle with its centre at point C.

The cut-outs around the holes at points E are 20 mm radius centred at points F.

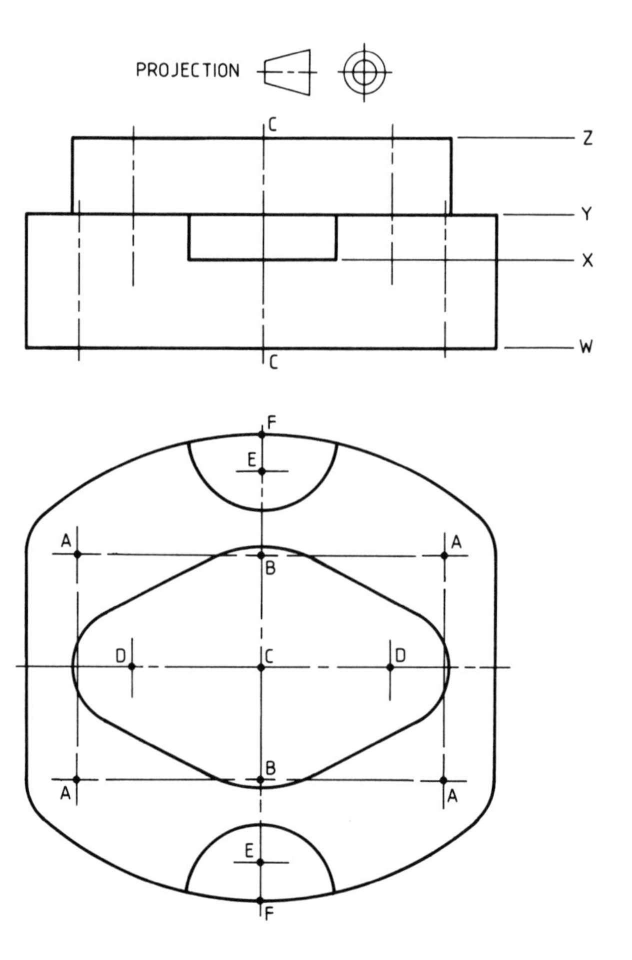

►**Figure 8.17** Spacer

Points A are at the corners of a rectangle 100 mm by 60 mm with its centre at point C. Points D and E are 35 mm and 52 mm from C, respectively.

Planes X, Y and Z are 22 mm, 35 mm and 55 mm, respectively, above plane W. Plane Z is the datum feature for the height dimensions.

All holes in the plan are positioned from point C.

Learning Assignment 9

Tolerances

> *A POINT TO NOTE*
> All dimensions, with the exception of auxiliary dimensions, are incomplete if a tolerance is not specified for them:
> - with the dimension; or
> - in a general note; or
> - in documentation associated with the drawing.

Dimensions specified for features of a part cannot be manufactured to exact sizes. Variations will occur because of inaccuracies inherent in the machine or manufacturing process being used, tool wear, temperature changes, differences in operator performance and so on. Therefore it is necessary to prescribe maximum and minimum sizes for the dimensions that will enable the finished part to function satisfactorily. The difference between the maximum and minimum sizes is the permitted variation for the dimension and is called the tolerance.

Tolerances applied to linear dimensions

The two ways in which these tolerances can be expressed on a drawing are illustrated in Figures 9.1 and 9.2.

In the first method, shown in Figure 9.1, the maximum and minimum sizes, or limits of size, are stated directly. The larger limit of size is placed above the smaller and both are given to the same number of decimal places.

The second method is shown in Figure 9.2. Here a size is specified for the dimension with limits of tolerance above and below that size. Both limits of tolerance have the same number of decimal places, except when one limit is zero. In this case the limit is written as '0', without a plus or minus sign. The dimension need not have the same number of decimal places as the limits of tolerance. The limit of tolerance which produces the larger limit of size is placed above that which produces the smaller limit of size.

The interpretations of the two methods are the same. They both do no more than define the maximum and minimum sizes for the dimension to which they are applied. However, stating the limits of size directly, as in Figure 9.1, avoids the operator having to calculate them from the dimension and limits of tolerance.

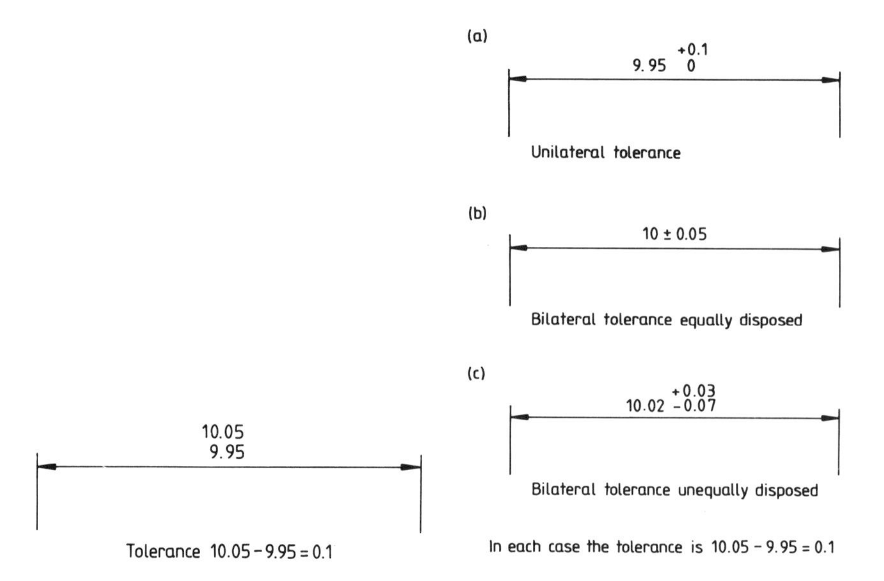

▲**Figure 9.1** Tolerancing – limits of size expressed directly

▲**Figure 9.2** Tolerancing – size specified with limits of tolerance

Unilateral and bilateral tolerances

In Figure 9.2(a) the application of the second method produces a unilateral tolerance. This means that variations in the size of the dimension are allowed in only one direction, and so one limit of tolerance is zero. Unilateral tolerances are advantageous when a critical size is approached as material is removed during manufacture, as with close-fitting holes and shafts.

Bilateral tolerances are also used. These allow variations in both directions in the size of the dimension, as shown in Figure 9.2(b) and (c). They are usually applied to location dimensions and any dimensions that can be allowed to vary in either direction.

> *A POINT TO NOTE*
>
> **BS 308 recommends that bilateral tolerances should preferably be disposed equally about the specified size of the dimension, as in Figure 9.2(b), rather than unequally as in Figure 9.2(c), except where this is required for the part to function satisfactorily.**

Tolerances applied to angular dimensions

The two methods used to tolerance linear dimensions can equally be applied to angular dimensions. Examples of toleranced angular dimensions are given in Figure 9.3.

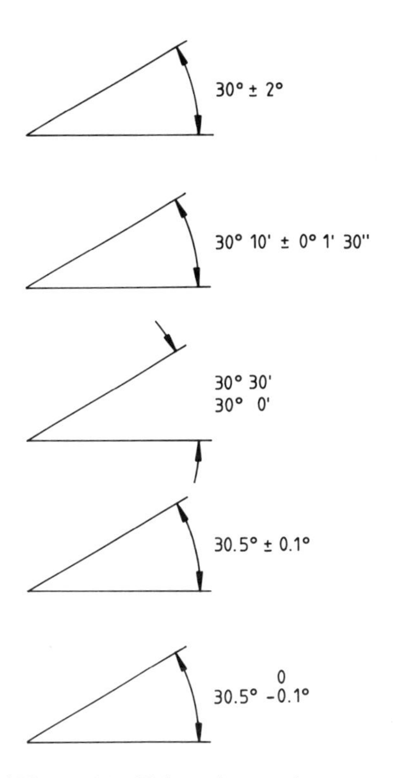

▲**Figure 9.3** Tolerancing angular dimensions

General tolerances

Where many dimensions on a drawing have the same, usually low, accuracy requirements a common tolerance can be applied to them. This is done by using a general note, so avoiding numerous repetitions of the same tolerance. The note is placed near the title block and examples which are often seen are:

- Tolerances unless otherwise stated:
- Linear ±0.2 Angular ±0° 30'.
- Tolerances on cast thicknesses: ±5%.
- For tolerances on forging dimensions see BS 4114.

Single limits of size

Where it is only necessary to specify one limit of size for a dimension, it should be followed by the abbreviation MAX or MIN. For example, the minimum length of full-form thread can be specified as 25 MIN FULL THD and the maximum radius permitted at a corner as R0.2 MAX. This means that the maximum full-form thread length and the minimum radius are uncontrolled. The radius, for example, could therefore be zero. Other applications of single limits of size include depths of holes, sizes of chamfers and so on.

Cumulative effects of tolerances

Where chain dimensioning of centre distances is used, the tolerances on the dimensions may add up through the chain, and this may be undesirable. In Figure 9.4(a) the tolerances on the distances from the edge of the plate to the hole axes are: to hole A ± 0.2; to hole B ± 0.4; and to hole C ± 0.6.

This accumulation may be avoided by using parallel dimensioning from a common datum feature instead of chain dimensioning. This is illustrated in Figure 9.4(b). The common datum feature is the plate edge, and now all three hole axes have a tolerance of ± 0.2 relative to it. However, note that the tolerance between any two holes is ± 0.4.

The drawing of the pivot pin in Figure 11.1 is an example of a part whose function requires the use of chain dimensioning.

A combination of chain and parallel dimensioning is required for most components.

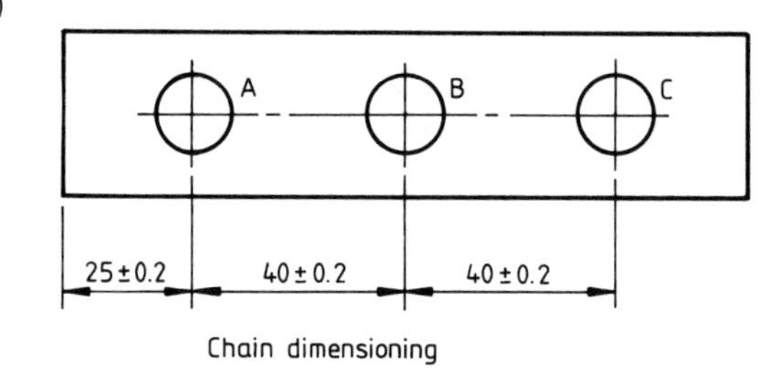

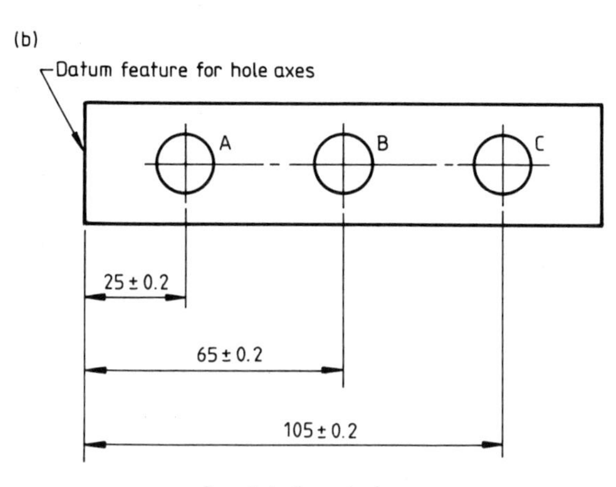

▶**Figure 9.4** Cumulative effect of tolerances

Additional information

Another way of preventing the tolerance accumulation inherent in chain dimensioning is to locate the hole axes with a positional tolerance, rather than by using toleranced linear dimensions. It is best to do this whenever more than two features need to be related together as a group. Positional tolerancing is an aspect of geometrical tolerancing, the subject of BS 308 : Part 3, and is outside of the scope of this workbook.

Size of tolerances

The largest tolerances which will provide satisfactory functioning of the part should be used. Specifying smaller tolerances than necessary will not produce a better part; it will make the part more expensive. The cost of production is directly related to the sizes of the tolerances. Small tolerances usually mean that the machine requires resetting more frequently because of tool wear; more skill is needed from the operator; and more time is necessary to make the part. Furthermore, small tolerances often mean that a particular machine or process has to be used, whereas if the tolerances were relaxed then it might be possible to use a simpler machine or cheaper process.

TASKS

Task 9.1

(a) What are the maximum and minimum lengths of the drilled plate shown in Figure 9.5?
(b) Redraw the plate using positive unilateral tolerances which will give a tolerance of 0.2 between any pair of holes and on the overall length.

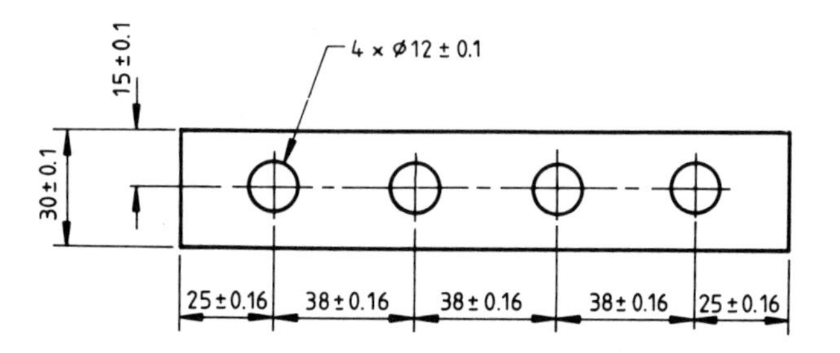

▶**Figure 9.5** Drilled plate

Task 9.2

Draw the given view of the end-plate in Figure 9.6 and dimension it to satisfy the following conditions:

(a) The five equally spaced bosses have their axes on an 80.0/80.1 mm diameter pitch circle. They have a diameter of 20 mm with a hole diameter of 10.00/10.08 mm. The blend radius between the bosses is 30 mm.
(b) The central hole is 22.00/22.05 mm diameter. The three slots are equally spaced with an angular tolerance of ± 10'. Their width is 4.000/4.012 mm and the diameter over their tops is 28 mm.
(c) The two small holes are 7.5/7.6 mm diameter and are spaced at 22.0 mm + 0.1 mm from the centre line of the plate.

Dimensions for which no tolerance is specified are to be subject to a general tolerance of ± 0.2 mm for linear dimensions and ± 30' for angles.

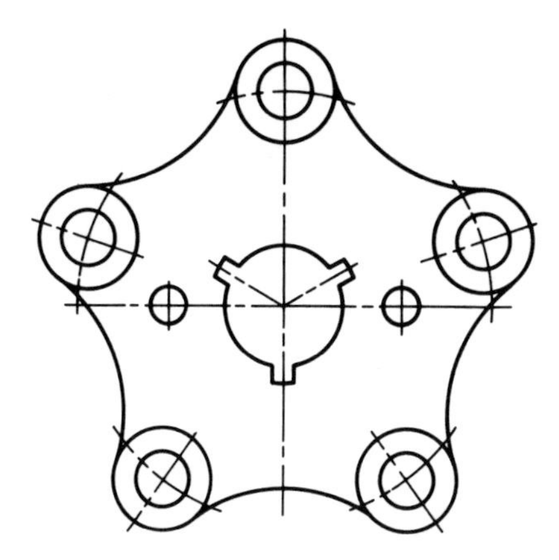

▲**Figure 9.6** End-plate

Learning Assignment 10

Limits and fits

The ISO system of limits and fits for engineering purposes is described in BS 4500. It provides a comprehensive range of limits and fits based on a series of tolerances graded to suit all classes of work from the finest to the coarsest. These tolerances are suitable for most general applications. They should be used whenever a graded series of tolerances is needed, whether the features concerned are members of a fit or not. The tolerances are not restricted solely to diameters, although only cylindrical parts – briefly designated 'holes' and 'shafts' – are referred to in the standard. They can also be applied, for example, to the width of a slot and the thickness of a key, and to lengths, heights and depths.

The system provides a very extensive range of fits, but in practice most engineering applications can be met with a small selection from the range. On the other hand, the system is very flexible and this means that special requirements, outside this restricted range, can be met without difficulty.

Additional information

The ISO system meets the two conditions which must be satisfied for a limits and fits system to be comprehensive; these are:

(a) Fits ranging from loose clearance to heavy interference and from coarse to fine quality must be available.
(b) Any fit in the system must have the same functional characteristics over the range of sizes covered by the system. For example, a precision running fit should ideally provide clearances which give precision running fit conditions whatever the size of the parts to be assembled.

A full explanation of the principles underlying the ISO system is given in BS 4500. It is outside the scope of this workbook and the interested reader should consult the standard.

Types of fit

A fit is the relationship between the sizes of the mating features of two assembled parts. The relationship determines which of the three types of fit – clearance, interference or transition – exists between the parts. Figure 10.1 shows the fits diagrammatically.

Clearance fits

A clearance fit, see Figure 10.1 (a), provides clearance whatever the sizes of the mating features are within their tolerances. Thus the shaft is always smaller than the hole into which it fits. Clearance fits are required for example for rotating shafts and their bores, bolts and their associated holes, and slides and slideways.

Interference fits

When the shaft is always larger than the hole then an interference fit exists between them as shown in Figure 10.1 (b). Some force is needed to assemble the members of an interference fit. Examples are for bushes pressed into bores, shrunk-on parts such as couplings and railway wheels and their shafts, and dowels and their fixing holes.

Transition fits

A transition fit (see Figure 10.1(c)), may provide either clearance or interference between the mating features. So the shaft may be smaller, larger or the same size as the hole into which it fits. Typical applications of transition fits are for spigots and their bores, and keys and their keyways.

Hole-basis and shaft-basis systems of fits

With the hole-basis system, illustrated in Figure 10.2(a), the hole diameter is kept constant and different fits are obtained by varying the shaft diameter. This system is in general use because it is usually easier to make the shaft (by turning and grinding) and measure it than the hole. So it is advantageous to be able to give most of the available tolerance to the hole and adjust the shaft size to suit the required fit. Furthermore, if the hole diameter is constant, fewer drills, reamers and gauges are needed, thus saving time and money.

Sometimes, however, it is better to use a shaft-basis system of fits, shown in Figure 10.2(b). This would be the case where a variety of parts such as bearings, couplings and collars have to be mounted on the same shaft. The shaft size is now kept constant and different fits are provided by varying the bore sizes of the parts which the shaft carries. But this in turn means that a series of tools and gauges is needed to make and check these bores, and this makes the system more expensive than the hole-basis system.

A simple example of the application of the shaft-basis system is shown in Figure 10.3. Here a hinged cover is attached to a casing. To retain the pivot pin and allow the cover to open, the pin is to be an interference fit in the lug on the casing and a clearance fit in the holes in the cover lugs. The pin therefore will have a constant diameter and to provide the required fits the diameters of the holes in the lugs must be varied.

A POINT TO NOTE

Shaft-basis fits are particularly appropriate when bar stock is available finished to standard shaft tolerances of the ISO system.

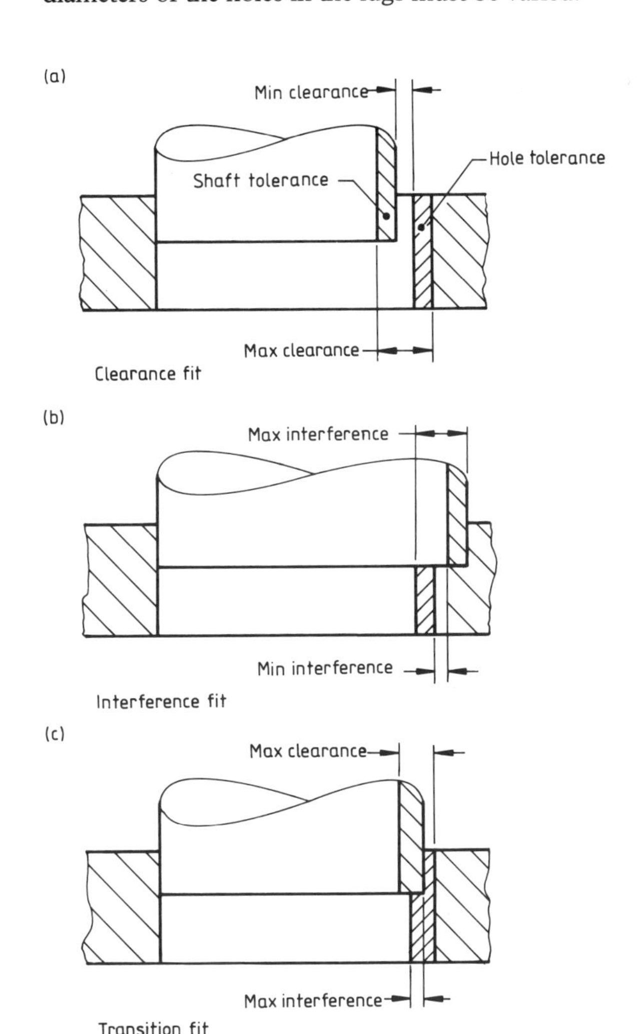

◀**Figure 10.1** Types of fit

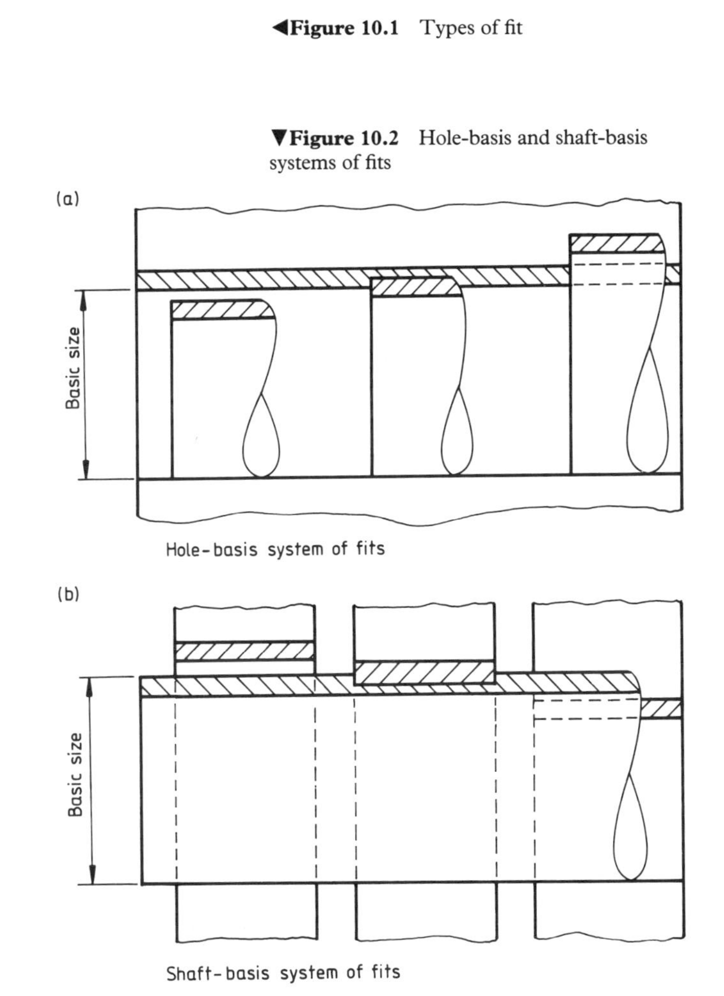

▼**Figure 10.2** Hole-basis and shaft-basis systems of fits

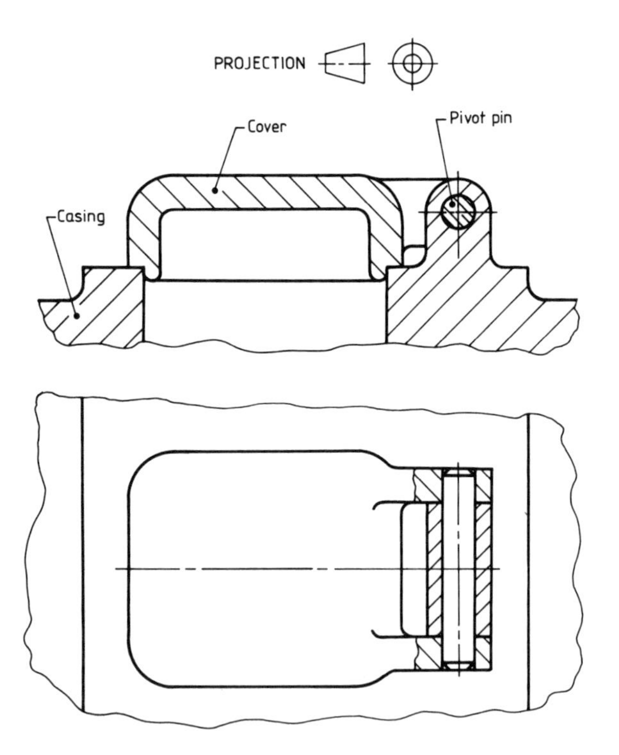

▶**Figure 10.3** Example of application of shaft-basis system of fits

BS 4500A *Selected ISO Fits – Hole Basis*

This data sheet, reproduced as Figure 10.4, is a small selection from the full range of fits given in BS 4500. It covers fits from loose clearance to heavy interference, and will be found sufficient for most applications. In fact, for the manufacture of some products, a further selection may be made from it.

The fits are derived from four hole tolerances (H11, H9, H8 and H7) and ten shaft tolerances (c11, d10, e9, f7, g6, h6, k6, n6, p6 and s6).

When using the data sheet bear in mind the following:

(a) Hole tolerances are designated by capital letters and shaft tolerances by lower-case letters. There are 27 letters in all, and they are used for both holes and shafts.

(b) The higher the number representing the tolerance grade, the bigger is the tolerance.

Only 'H' holes have been selected for inclusion in the data sheet. The five selected tolerance grades are taken from a total of 18.

The limits of size for a feature are defined by the basic size of the feature, say 35 mm, followed by the tolerance designation, for example 35H8 or 35f7.

The basic size is the same for both members of a fit; in the data sheet it is called the nominal size. For all 'H' holes one limit of size is the basic size.

A fit is indicated by combining the basic size with the designations for both members. The designation of the hole limits is always given first, thus: 35H8 – f7 or 35H8/f7.

Classifications and typical applications of the fits in BS 4500A

The fits in BS 4500 have not been given names. The classifications which follow are generally accepted, but variations do exist. For example, an H11–c11 fit may be classified as a loose clearance fit, and an H7–p6 fit as a press or light press fit.

The fits are intended to be used for parts made from ordinary engineering materials operating under normal conditions. In addition, in dealing with very large sizes, several factors, including the effects of temperature, must be taken into account. These factors are discussed in BS 4500.

Extracted from BS 4500 : 1969

BRITISH STANDARD

SELECTED ISO FITS—HOLE BASIS

Data Sheet
4500A
Issue 1. February 1970
Confirmed August 1985

Diagram to scale for 25 mm. diameter: H11/c11, H9/d10, H9/e9, H8/f7, H7/g6, H7/h6, H7/k6, H7/n6, H7/p6, H7/s6 (Holes, Shafts)

Nominal sizes		Clearance fits												Transition fits				Interference fits				Nominal sizes	
		Tolerance		Tolerance		Tolerance		Tolerance		Tolerance		Tolerance		Tolerance		Tolerance		Tolerance		Tolerance			
Over	To	H11	c11	H9	d10	H9	e9	H8	f7	H7	g6	H7	h6	H7	k6	H7	n6	H7	p6	H7	s6	Over	To
mm	mm	0·001 mm	0·001 mm	0·001 mm	0·001 mm	0·001 mm	0·001 mm	0·001 mm	0·001 mm	0·001 mm	0·001 mm	0·001 mm	0·001 mm	0·001 mm	0·001 mm	0·001 mm	0·001 mm	0·001 mm	0·001 mm	0·001 mm	0·001 mm	mm	mm
—	3	+ 60 0	− 60 − 120	+ 25 0	− 20 − 60	+ 25 0	− 14 − 39	+ 14 0	− 6 − 16	+ 10 0	− 2 − 8	+ 10 0	− 6 0	+ 10 0	+ 6 + 0	+ 10 0	+ 10 + 4	+ 10 0	+ 12 + 6	+ 10 0	+ 20 + 14	—	3
3	6	+ 75 0	− 70 − 145	+ 30 0	− 30 − 78	+ 30 0	− 20 − 50	+ 18 0	− 10 − 22	+ 12 0	− 4 − 12	+ 12 0	− 8 0	+ 12 0	+ 9 + 1	+ 12 0	+ 16 + 8	+ 12 0	+ 20 + 12	+ 12 0	+ 27 + 19	3	6
6	10	+ 90 0	− 80 − 170	+ 36 0	− 40 − 98	+ 36 0	− 25 − 61	+ 22 0	− 13 − 28	+ 15 0	− 5 − 14	+ 15 0	− 9 0	+ 15 0	+ 10 + 1	+ 15 0	+ 19 + 10	+ 15 0	+ 24 + 15	+ 15 0	+ 32 + 23	6	10
10	18	+ 110 0	− 95 − 205	+ 43 0	− 50 − 120	+ 43 0	− 32 − 75	+ 27 0	− 16 − 34	+ 18 0	− 6 − 17	+ 18 0	− 11 0	+ 18 0	+ 12 + 1	+ 18 0	+ 23 + 12	+ 18 0	+ 29 + 18	+ 18 0	+ 39 + 28	10	18
18	30	+ 130 0	− 110 − 240	+ 52 0	− 65 − 149	+ 52 0	− 40 − 92	+ 33 0	− 20 − 41	+ 21 0	− 7 − 20	+ 21 0	− 13 0	+ 21 0	+ 15 + 2	+ 21 0	+ 28 + 15	+ 21 0	+ 35 + 22	+ 21 0	+ 48 + 35	18	30
30	40	+ 160 0	− 120 − 280	+ 62 0	− 80 − 180	+ 62 0	− 50 − 112	+ 39 0	− 25 − 50	+ 25 0	− 9 − 25	+ 25 0	− 16 0	+ 25 0	+ 18 + 2	+ 25 0	+ 33 + 17	+ 25 0	+ 42 + 26	+ 25 0	+ 59 + 43	30	40
40	50	+ 160 0	− 130 − 290																			40	50
50	65	+ 190 0	− 140 − 330	+ 74 0	− 100 − 220	+ 74 0	− 60 − 134	+ 46 0	− 30 − 60	+ 30 0	− 10 − 29	+ 30 0	− 19 0	+ 30 0	+ 21 + 2	+ 30 0	+ 39 + 20	+ 30 0	+ 51 + 32	+ 30 0	+ 72 + 53	50	65
65	80	+ 190 0	− 150 − 340																	+ 30 0	+ 78 + 59	65	80
80	100	+ 220 0	− 170 − 390	+ 87 0	− 120 − 260	+ 87 0	− 72 − 159	+ 54 0	− 36 − 71	+ 35 0	− 12 − 34	+ 35 0	− 22 0	+ 35 0	+ 25 + 3	+ 35 0	+ 45 + 23	+ 35 0	+ 59 + 37	+ 35 0	+ 93 + 71	80	100
100	120	+ 220 0	− 180 − 400																	+ 35 0	+ 101 + 79	100	120
120	140	+ 250 0	− 200 − 450	+ 100 0	145 305	+ 100 0	− 84 − 185	+ 63 0	− 43 − 83	+ 40 0	− 14 − 39	+ 40 0	− 25 0	+ 40 0	+ 28 + 3	+ 40 0	+ 52 + 27	+ 40 0	+ 68 + 43	+ 40 0	+ 117 + 92	120	140
140	160	+ 250 0	− 210 460																	+ 40 0	+ 125 + 100	140	160
160	180	+ 250 0	− 230 − 480																	+ 40 0	+ 133 + 108	160	180
180	200	+ 290 0	− 240 − 530	+ 115 0	− 170 − 355	+ 115 0	− 100 − 215	+ 72 0	− 50 − 96	+ 46 0	− 15 − 44	+ 46 0	− 29 0	+ 46 0	+ 33 + 4	+ 46 0	+ 60 + 31	+ 46 0	+ 79 + 50	+ 46 0	+ 151 + 122	180	200
200	225	+ 290 0	260 550																	+ 46 0	+ 159 + 130	200	225
225	250	+ 290 0	− 280 − 570																	+ 46 0	+ 169 + 140	225	250
250	280	+ 320 0	− 300 − 620	+ 130 0	− 190 − 400	+ 130 0	− 110 − 240	+ 81 0	− 56 − 108	+ 52 0	− 17 − 49	+ 52 0	32 0	+ 52 0	+ 36 + 4	+ 52 0	+ 66 + 34	+ 52 0	+ 88 + 56	+ 52 0	+ 190 + 158	250	280
280	315	+ 320 0	330 − 650																	+ 52 0	+ 202 + 170	280	315
315	355	+ 360 0	360 − 720	+ 140 0	− 210 − 440	+ 140 0	− 125 − 265	+ 89 0	− 62 − 119	+ 57 0	− 18 − 54	+ 57 0	− 36 0	+ 57 0	+ 40 + 4	+ 57 0	+ 73 + 37	+ 57 0	+ 98 + 62	+ 57 0	+ 226 + 190	315	355
355	400	+ 360 0	400 760																	+ 57 0	+ 244 + 208	355	400
400	450	+ 400 0	− 440 840	+ 155 0	− 230 − 480	+ 155 0	− 135 − 290	+ 97 0	− 68 − 131	+ 63 0	− 20 − 60	+ 63 0	− 40 0	+ 63 0	+ 45 + 5	+ 63 0	+ 80 + 40	+ 63 0	+ 108 + 68	+ 63 0	+ 272 + 232	400	450
450	500	+ 400 0	480 880																	+ 63 0	+ 292 + 252	450	500

BRITISH STANDARDS INSTITUTION, 2 Park Street, London, W1A 2BS
SBN: 580 05766 6

▲Figure 10.4 BS 4500A *Selected ISO Fits – Hole Basis*

H11–c11 Extra loose running fit

Provides a large clearance. Useful in dirty conditions and where corrosion may be present, as for example, with agricultural machinery. Gives easy assembly or a close fit at high working temperatures, for example for internal-combustion exhaust valves in their guides.

H9–d10 Loose running fit

Used for parts which need ease of assembly such as loose pulleys, large bearings and gland seals.

H9–e9 Easy running fit

Suitable for smaller parts which require fairly large clearances. Used where a shaft is supported by widely separated bearings or by several bearings in line. Examples are camshafts, rocker shafts and gear-box selector shafts.

H8–f7 Normal running fit

In general use to provide good-quality fits which can be easily produced. Common applications are for gears running on fixed shafts and bearings for gear-box shafts.

H7–g6 Close running or location fit

Is expensive to produce because of the small clearance. Is unsuitable for continuous running, except under very light loads. Can be used for precision sliding fits between flat surfaces, oscillating bearings for linkage pins and for easy-location fits such as fitted bolts in a bearing cap.

H7–h6 Precision slide or location fit

The clearance provided by this fit is on average very small, which makes it suitable for precision sliding fits with good lubrication which run under constant-temperature conditions. Can also be used for oscillating fits such as that for steel gudgeon pins in bronze small-end bearings, and for non-running assemblies such as jig location fits and exhaust valve guides in internal-combustion-engine cylinder heads.

H7–k6 Push or easy keying fit – transition

Suitable for location fits which are not often dismantled, or where vibration between the parts cannot be allowed. Examples are parts on rotating machinery such as gear rings on hubs and couplings on shafts.

H7–n6 Tight keying fit – transition

Used for location fits when the amount of clearance which can occur with an H7–k6 fit is too great. On long parts has the effect of an interference fit. Ensures tight assembly of lightweight rotating parts such as fan blades and commutator shells.

H7–p6 Light drive fit – interference

Used to provide a secure fixing for steel parts which are to be dismantled and renewed in service. The light interference avoids damage or excessive stress to the parts during assembly and removal.

H7–s6 Heavy drive fit – interference

This fit provides permanent assembly for ferrous parts below about 25 mm diameter and semi-permanent assembly for larger sizes. These may need to be shrunk-fit to avoid surface damage during assembly. Can be used to fit crank pins to cast-iron crankshaft webs, cylinder liners into internal-combustion-engine cylinder blocks and bronze bushes into alloy castings.

Additional information

Data sheet BS 4500B provides a similar selection of fits to those in BS 4500A but on a shaft basis.

Examples of the use of BS 4500A

Example 1

To find the limits of size for the members of the fit 35H8–f7 from BS 4500A proceed as follows:

- Find the basic size of 35 mm in the nominal sizes columns. Read across to the column headed H8 and obtain the tolerance, +39/0. The tolerances are given in micrometres (0.001 mm), so the limits of size for the hole are:

  35.000 + 0.039 = 35.039 mm diameter maximum and
  35.000 + 0 = 35.000 mm diameter minimum.

Using the same procedure for the 35f7 shaft the limits of size are:

35.000 – 0.025 = 34.975 mm diameter maximum and
35.000 – 0.050 = 34.950 mm diameter minimum.

POINTS TO NOTE

- **The nominal sizes column 'Over' means 'over but excluding', and the column headed 'To' means 'up to and including'.**
- **Remember that the limits of size for a feature must be given on the drawing to the same number of decimal places.**

Example 2

Figure 10.5 shows a dowel pin assembly with the required fits. Find the limits of size for the dowel and the mating holes.

For the interference fit:

- The hole diameter is:

  12.000 + 0.018 = 12.018 mm maximum and
  12.000 + 0 = 12.000 mm minimum.

- The dowel diameter is:

  12.000 + 0.029 = 12.029 mm maximum and
  12.000 + 0.018 = 12.018 mm minimum.

For the clearance fit:

- The hole diameter is:

  10.000 + 0.015 = 10.015 mm maximum and
  10.000 + 0 = 10.000 mm minimum.

- The dowel diameter is:

  10.000 – 0 = 10.000 mm maximum and
  10.000 – 0.009 = 9.991 mm minimum.

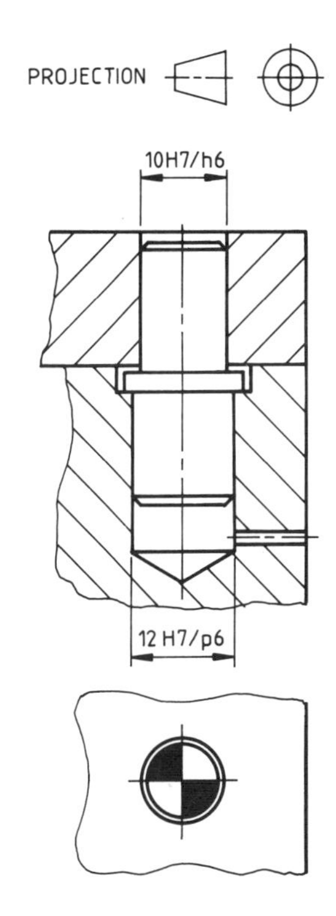

▲**Figure 10.5** Dowel assembly

Additional information

The dowel in Figure 10.5 on page 57 is an interference fit in the blind hole in the lower part. (A blind hole does not go right through the part.) The small hole drilled into the blind hole allows air trapped below the dowel to escape. If the air cannot escape then it will be compressed to such an extent that assembly will be impossible. The simplest way of ensuring assembly is for the hole for the dowel to go right through, providing that the function of the part will allow this. Alternatively, a small hole can be drilled axially through the dowel, or a small flat can be machined on the interference fit diameter.

Figure 10.5 illustrates the convention of identifying a dowel by filling-in opposite quadrants of the circular view. This is very often done, although it is not recommended in BS 308.

TASKS

Task 10.1

For the locating pin assembly shown in Figure 10.6, determine the limits of size for the features forming the three fits. Make dimensioned drawings of the pin and pin bush.

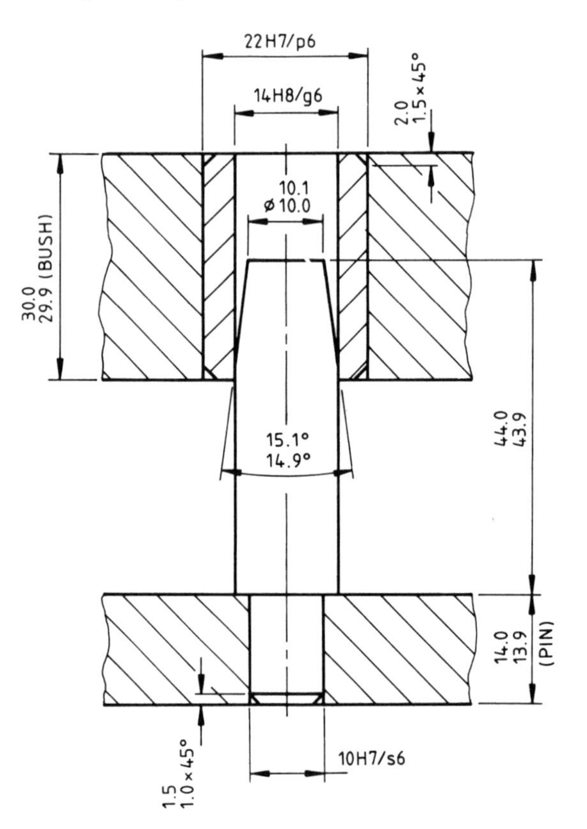

▶**Figure 10.6** Locating pin assembly

Additional information

In Figure 10.6 the use of chamfers on the pin and bush to ease the assembly of the interference fits is a common design feature.

When a bush is assembled with an interference fit then the bore may partly close. To avoid this the bore may be finished by reaming after assembly.

TASKS

Task 10.2

In the idler gear assembly shown in Figure 10.7 all the holes and the spigot recess are to have H7 tolerances. Select suitable fits for the spigot, shaft and bearing using the basic sizes shown, the bearing to be an interference fit in the bore of the gear. Find the limits of size for the pair of features in each fit. Treat the bearing length and gear thickness as a hole and shaft fit. Make a dimensioned drawing of the shaft bearing.

The abbreviation 'BSC' on Figure 10.7 means 'basic'.

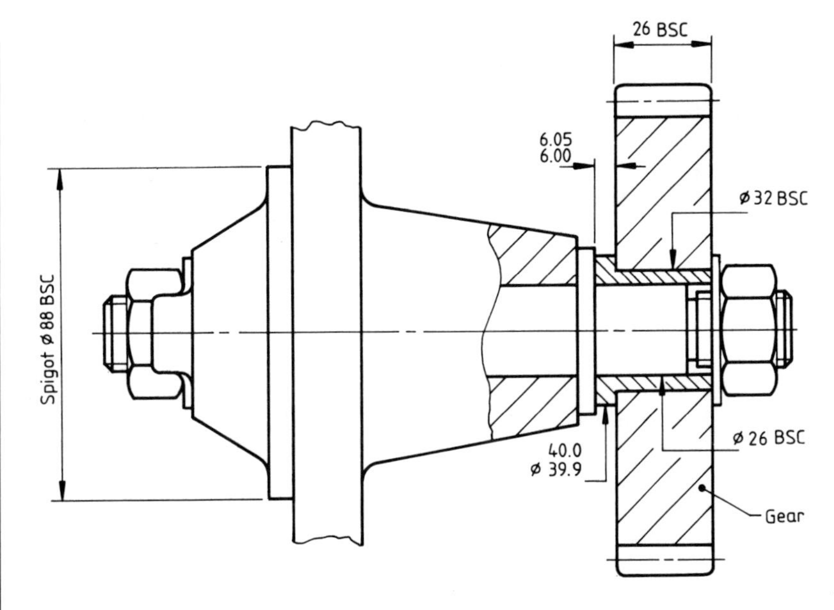

Figure 10.7 Idler gear assembly

Task 10.3

Figure 10.8 shows part of the design scheme for a turntable, the items for which are arranged as follows.

The turntable (item 2) rotates on the centre-piece (item 1), which is located in the cross-piece (item 3), by a spigot. The centre-piece is retained by an M20 bolt (item 4) and plain washer (item 5).

Select suitable fits between items 1 and 2, and items 1 and 3 and make a fully dimensioned drawing of item 1. All dimensions are to carry suitable tolerances. Before settling the tolerance for the clearance hole for the bolt, consult a fastener manufacturer's catalogue or BS 3692 to find the limits of size for the bolt diameter.

Make any necessary modifications to the scheme which will assist production and assembly.

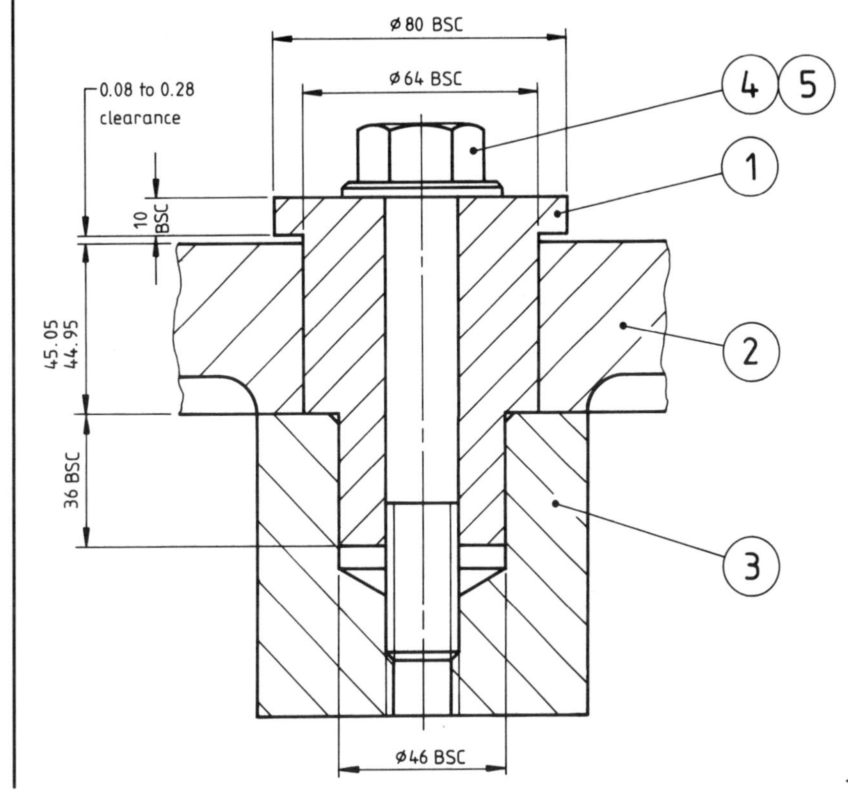

Figure 10.8 Turntable assembly

Learning Assignment 11

Selection of dimensions

All dimensions and other information necessary for the complete size description and satisfactory functioning of an object must be shown directly on the drawing and appear once only. If a dimension is duplicated and then has to be changed, one entry may be overlooked. The same dimension will then appear with two values and confusion will result. By the same token, superfluous dimensions must not be given, or a size will be fixed in two different ways. Bear in mind that the dimensions used to make the drawing are not necessarily those used to make the part.

The dimensions must define the condition in which the part is to be used and include any necessary surface treatment or finishing except painting or lacquering.

Do not specify production processes or inspection methods unless they are essential to ensure satisfactory functioning or interchangeability.

Size and location dimensions

In the main, engineering parts are made up of simple geometrical shapes, such as cylinders and prisms. These shapes result from the need to keep the forms of parts as simple as possible for economy of manufacture, and from the fundamental manufacturing operations. Exterior shapes such as shafts may be thought of as positive; holes, counterbores and similar interior shapes as negative.

The first step in dimensioning any engineering part is to give the sizes of its simple geometrical shapes. These are the size dimensions and methods of applying them to common features are given in Assignment 8. Secondly, the dimensions which locate the shapes relative to each other – the location dimensions – are added.

Prisms are located by their surfaces; cylinders, tapers and similar features by their axes. Castings and forgings are only partially machined and location dimensions for their features must be taken from the machined surfaces or the axes of machined bores.

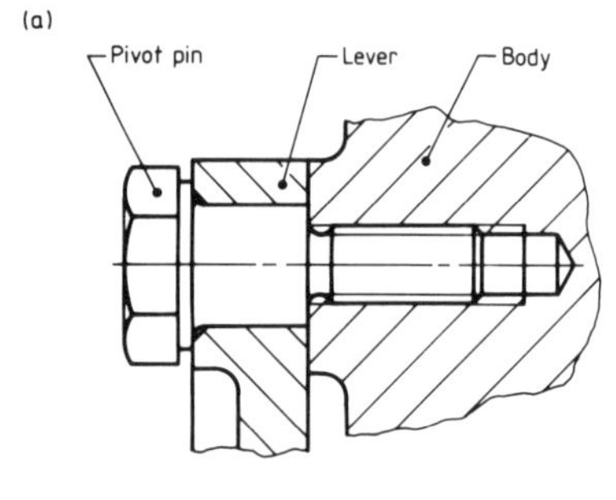

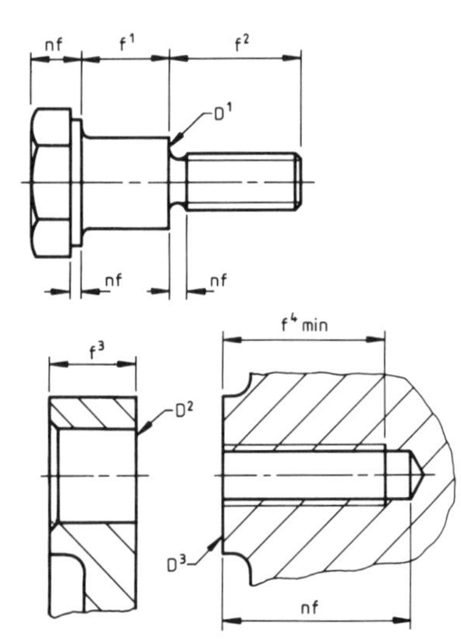

▲Figure 11.1 Functional dimensions

Additional information

If a CAM program is to be used to machine a hole whose axis is closely controlled for position, then the resolution of position chosen on the CAD system will need to be accurate enough to achieve the specified positional tolerance for the axis.

Functional dimensions

Some dimensions of a part are essential for it to function satisfactorily, and these functional dimensions must be stated directly on the drawing and not have to be calculated or inferred. In Figure 11.1(a) for example, the functional requirement for the assembly is that the lever should be able to swing on the pivot pin. This means that the thickness of the boss on the lever must be slightly less than the length of the bearing surface on the pivot pin. In addition, to ensure that the shoulder on the pin will be tightened against the face on the body, the thread on the pin must not reach the end of the thread in the body.

From these considerations the functional length dimensions are f^1 to f^4 as shown in Figure 11.1(b), and the datum features for them, D^1, D^2 and D^3, can be deduced. Dimensions f^1 and f^3 have the same basic size, but with appropriate tolerances the required functional clearance can be obtained as explained below. The length f^4 of the thread in the body need only be given as a minimum.

The remaining length dimensions, marked 'nf', are non-functional and their arrangement may be chosen to assist the manufacture or inspection of the part.

If a datum feature which is convenient for measurement is chosen instead of the correct one for functioning then smaller tolerances will almost always be necessary. This may mean that parts which would satisfy the functional requirements of the assembly will be rejected because they exceed those smaller tolerances.

A POINT TO NOTE

When parts are to be made using a CNC machining system, the drawings are generally dimensioned entirely from a common datum point instead of from the functional datum features. However, the accuracy of these systems is such that the smaller tolerances which result can usually be met.

Tolerances on functional dimensions

Functional requirements are often toleranced gaps or clearances in assemblies where several parts have a bearing on the function. An elementary example is shown in Figure 11.2. Here the functional requirement F – a controlled clearance – is affected by the three parts A, B and C and their functional dimensions a, b and c. The sum of the tolerances on the functional dimensions must equal the tolerance on the functional requirement.

Suppose the functional requirement is to be 0.2 mm to 0.5 mm, and the basic sizes of the functional dimensions are a = 30 mm, b = 20 mm and c = 10 mm. Then the dimension equation is

$$F = 30 - 20 - 10$$

and if we take the minimum clearance of 0.2 mm, then

$$0.2 = 30 - 20 - 10 + 0.2$$

The 0.2 on the right-hand side is needed to balance the equation. It can be eliminated by adjusting one of the functional dimensions appropriately, say by reducing the 10 mm dimension by 0.2 mm. Then the equation becomes

$$0.2 = 30 - 20 - 9.8 \quad (1)$$

The tolerance on the functional requirement is 0.5 – 0.2 = 0.3 mm. This must be distributed among the functional dimensions, say as 0.14 mm to the 30 mm dimension and 0.08 mm each to the 20 mm and 9.8 mm dimensions. This gives the tolerance distribution equation

$$0.3 = 0.14 + 0.08 + 0.08 \quad (2)$$

By combining equations (1) and (2) we can find the limits of tolerance for the functional dimensions. Thus

$$0.2^{+0.3}_{0} = 30^{+0.14}_{0} - 20^{0}_{-0.08} - 9.8^{0}_{-0.08}$$

The functional dimensions expressed as limits of size are a = 30.14/30.00, b = 20.00/19.92, c = 9.80/9.72.

As a check, use these values to calculate the maximum and minimum values of the functional requirement. The maximum value is given when a is a maximum and b and c are minima. The other values of a, b and c give the minimum value:

$$30.14 - 19.92 - 9.72 = 0.5 \text{ and}$$
$$30.00 - 20.00 - 9.80 = 0.2$$

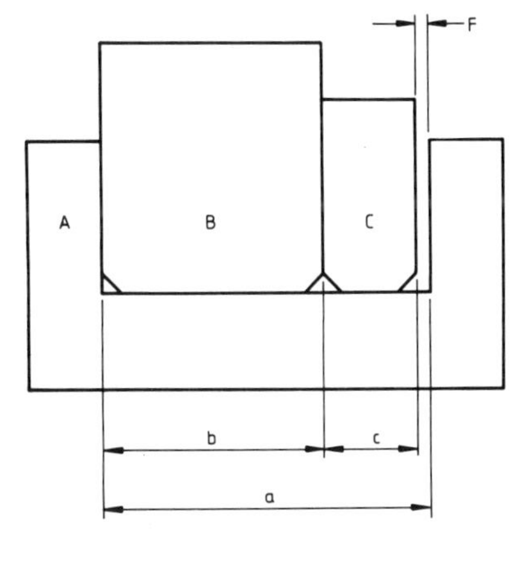

▲Figure 11.2 Assembly with functional requirement and dimensions

POINTS TO NOTE

- A part has only one functional dimension related to any one functional requirement.
- When allocating tolerances to functional dimensions, give more of the functional requirement tolerance to the dimensions that are difficult to control. In Figure 11.2 it is probably more difficult to control and measure the gap in part A than the thicknesses of the plates B and C.
- The signs of the limits of tolerance for functional dimensions are found as follows:
  - (a) When a functional dimension and the functional requirement have the same sign, the signs of their limits of tolerance are the same. Therefore

    $$(+)12^{+0.1}_{0} = (+)8^{+0.06}_{0} + 4^{+0.04}_{0}$$

    $$\text{or } (+)12^{0}_{-0.1} = (+)8^{0}_{-0.06} + 4^{0}_{-0.04}$$

  - (b) When the signs of functional requirement and functional dimension are opposite, their limits of tolerance have opposite signs. Therefore

    $$(+)7^{+0.1}_{0} = (+)13^{+0.06}_{0} - 6^{0}_{-0.04}$$

    $$\text{or } (+)7^{0}_{-0.01} = (+)13^{0}_{-0.06} - 6^{+0.04}_{0}$$

Task 11.1

Figure 11.3 shows views of a pulley assembly. Prepare fully dimensioned drawings of the pulley and base, selecting suitable 'H' holes where necessary.

To allow the roller to rotate on the pin its end-float between the bosses on the base is to be 0.1 mm minimum to 0.15 mm maximum.

Add a general note to the drawings stating the values of unspecified linear and angular tolerances.

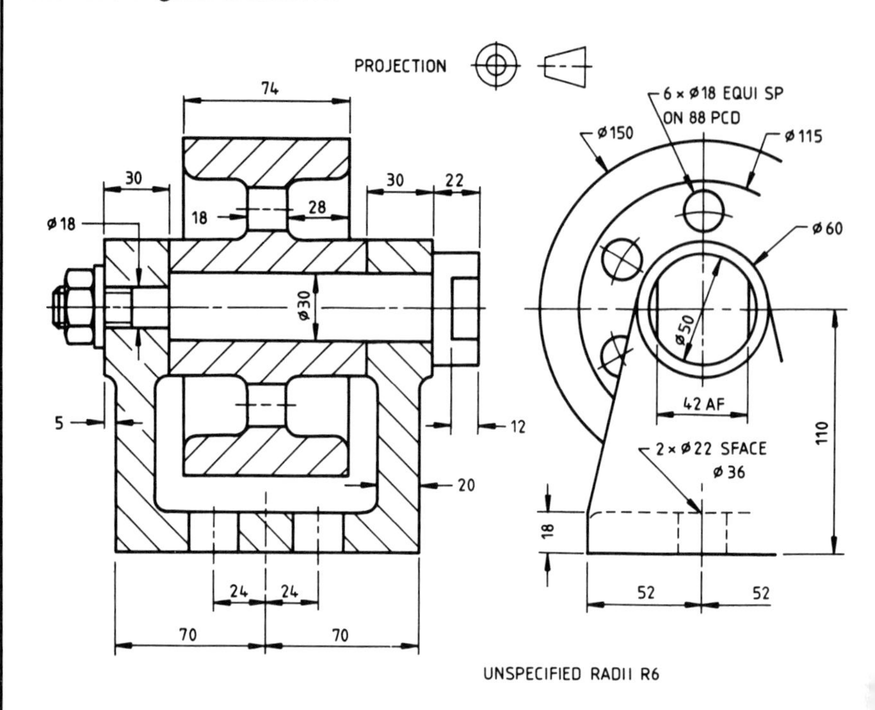

▶**Figure 11.3** Roller assembly

Task 11.2

In Figure 11.4 the end-float of the shaft collar between the bush flange and the cover spigot is to be 0.25 mm to 0.5 mm. Make fully dimensioned drawings for the shaft end, bush and cover, selecting suitable fits for the cover spigot in the body, the shaft in the bush and the bush in the body. Also draw a view of the body sufficient to show only the functional dimension for this part and the dimensioned mating diameter for the cover spigot.

Any necessary modifications to the parts to ease manufacture or assembly must be shown. State general tolerance values for untoleranced dimensions.

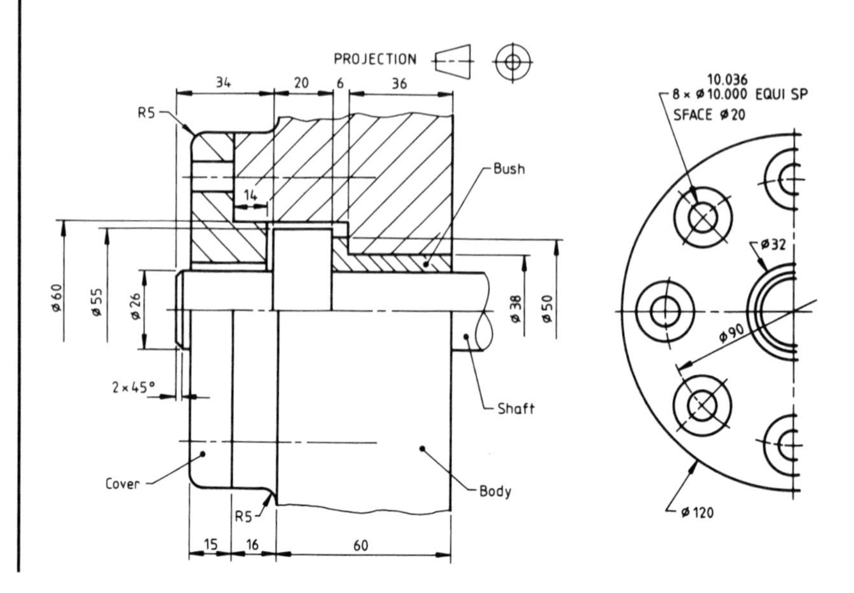

▶**Figure 11.4** Cover assembly

Learning Assignment 12

Conventional representation of screw threads

The crests and roots of a screw thread are curves and to draw them correctly is very time-consuming. So to reduce the amount of work a conventional representation is used, and this is the same whichever type or size of screw thread is being drawn.

Figure 12.1 illustrates the terms associated with screw threads.

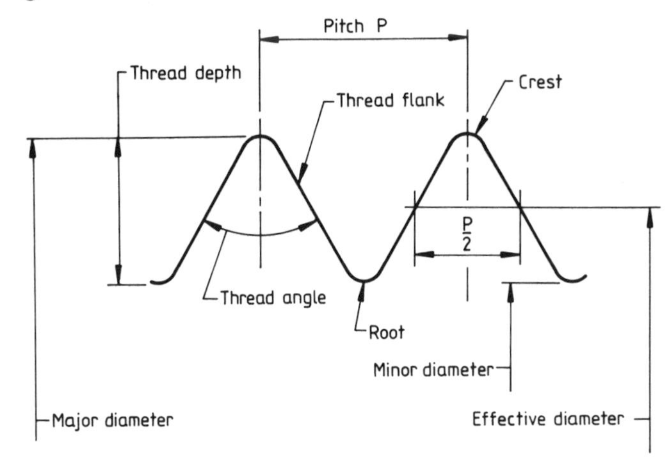

◀**Figure 12.1** Screw thread terms

External threads

The conventional representation of external threads is given in Figure 12.2. Figure 12.2(a) shows how they appear in an outside view. The major diameter, which is the diameter over the crests of the thread, is drawn using

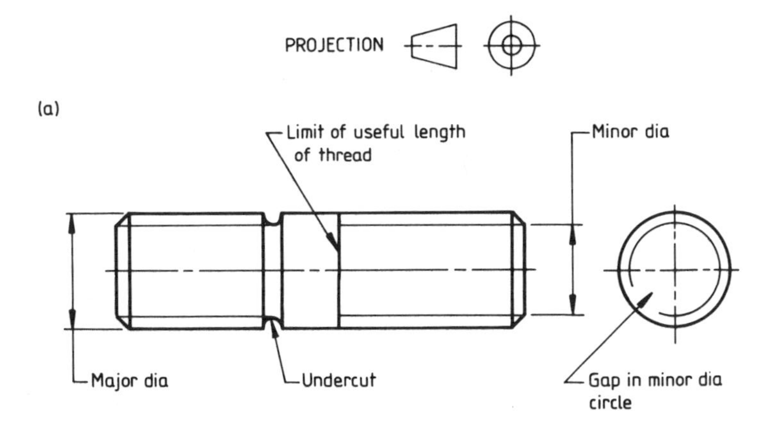

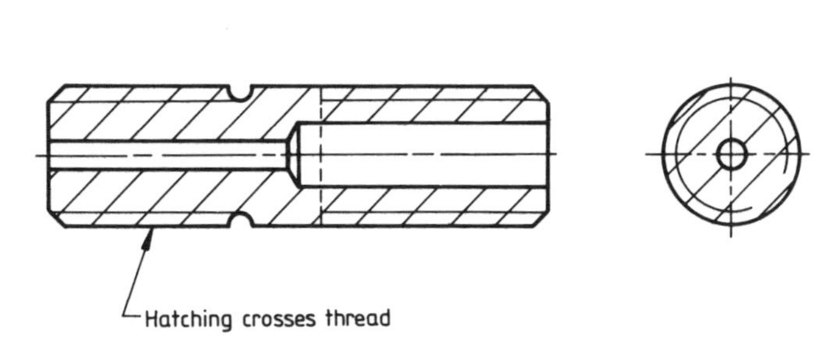

◀**Figure 12.2** Conventional representation of external screw threads

a pair of thick lines. The minor diameter, which is the diameter over the roots of the thread, is represented by a pair of thin lines. The limit of useful length of thread, that is the point at which the root of the thread ceases to be fully formed, is shown by a thick line. At this point the thread run-out begins and the thread becomes progressively shallower until it disappears. On the circular view of the thread the major diameter appears as a thick circle and the minor diameter as a thin circle with a gap.

In Figure 12.2(b) the conventional representation of external threads in section is shown. The minor diameter, as before, is drawn with a pair of thin lines. These are crossed by the hatching, which stops at the major diameter.

TASKS

Task 12.1

The pin in Figure 12.3 is to have a screw thread 50 mm long machined where shown. Draw the given view as an undimensioned half section, showing the thread.

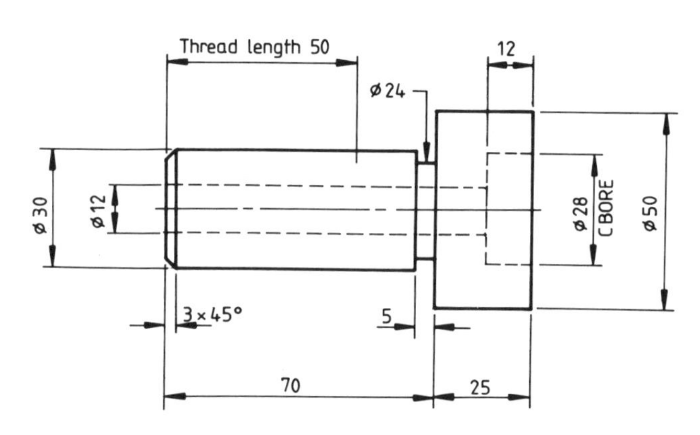

▶**Figure 12.3** Threaded pin

Internal threads

The major diameter of an internal thread is the diameter over the roots of the thread and the minor diameter is the diameter over the crests.

When an outside view of an internal thread is drawn it must be represented entirely by dashed lines, as shown in Figure 12.4(a). Figure 12.4(b), (c) and (d) give examples of internal threads in section. In Figure 12.4(b) and (c) the threads pass only part way through the part, and are therefore called blind tappings. A through tapping, in which the thread passes right through the part, is illustrated in Figure 12.4(d).

In the sectional view of an internal thread the minor diameter drilling, which is produced by the tapping drill, is drawn with thick lines. A thick line is also used for the line at the end of the full-form thread in a blind tapping. The hatching crosses the thread, in the same way as for a sectioned external thread. In the circular view the minor diameter is shown as a complete thick circle and the major diameter as a thin circle with a gap.

POINTS TO NOTE

- It is important that thick and thin lines are used in their correct places.
- Although BS 308 recommends that the thread depth is drawn to scale, it is usually sufficient for the major diameter to be drawn at its correct size with the thread depth roughly in the proportion shown in the figures here.
- The included angle of the cone produced by the point of the tapping drill is 120°, not 90°. Do not omit the line between this cone and the cylindrical part of the drilled hole.

Additional information

An internal thread is known as a tapping because it is often produced by a tool called a tap.

The tapping drill depth is almost always greater than the useful thread length. This prevents the tap jamming in the bottom of the hole and breaking when the thread is being formed. The hole is said to be bottomed tapped when the useful thread length and tapping drill depth are the same.

The 45° chamfer at the beginning of an external thread removes the first thin sharp thread for safety.

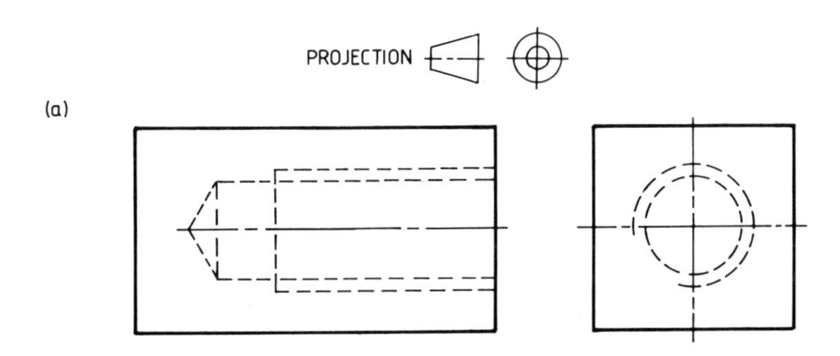

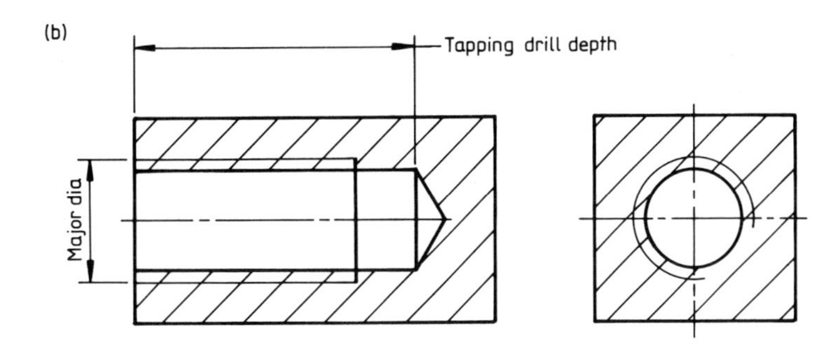

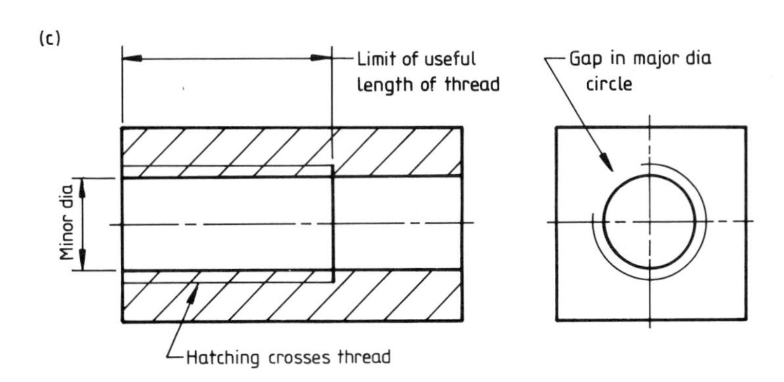

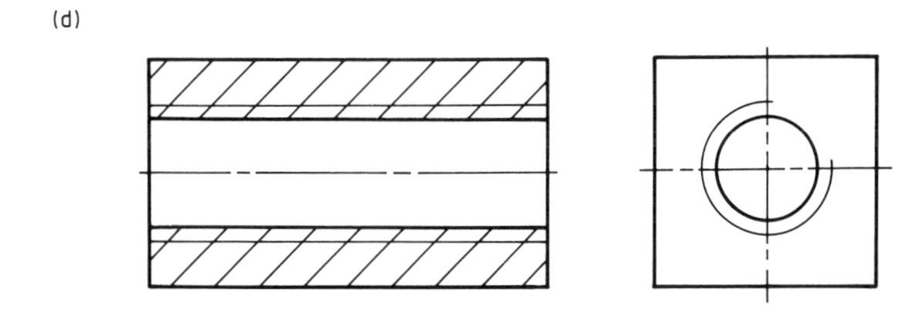

(a), (b) and (c) are blind tappings; (d) is a through tapping

◀**Figure 12.4** Conventional representation of internal screw threads

TASKS

Task 12.2

Draw without dimensions Figure 12.5, which represents the thickness of a piece of tapped plate. Show in section:

- At A a through tapping with a major diameter of 16 mm. Both ends of the tapping are to be countersunk to 22 mm diameter at 90°.
- At B a blind tapping with a major diameter of 20 mm, the thread depth being 35 mm, with the tapping drill going right through the plate.
- At C a blind tapping with a major diameter of 18 mm, a thread depth of 30 mm and a tapping drill depth of 35 mm. The end of the tapping is to be counterbored to 25 mm diameter for a depth of 6 mm.

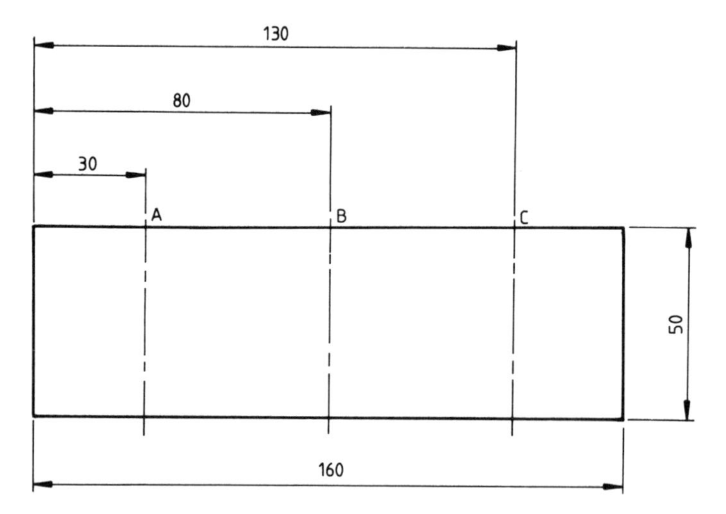

▶**Figure 12.5** Tapped plate

Assembled threads

When assembled threads are drawn, external threads are shown covering internal threads, as in Figure 12.6. In Figure 12.6(a) the screw with the external thread is – by convention – unsectioned, and so thick lines are used for its major diameter and thin lines for the minor diameter. The opposite applies to the sectioned thread in the tapping and the hatching crosses this thread.

When the external thread is sectioned, as in Figure 12.6(b), its minor diameter is drawn with thin lines which are crossed by the hatching. But note that the external threads are again shown covering the internal threads.

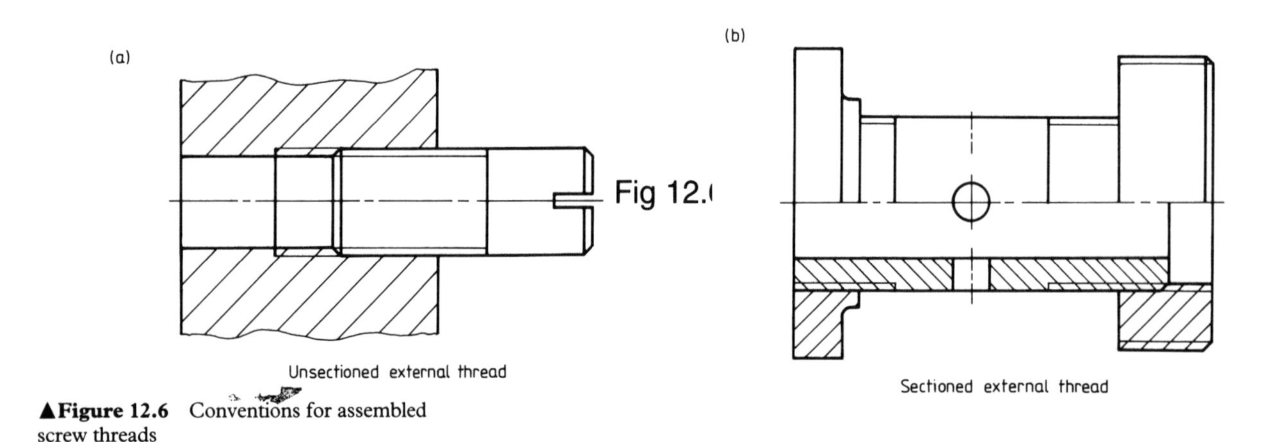

▲**Figure 12.6** Conventions for assembled screw threads

Task 12.3

Figure 12.7 shows the parts for a cover assembly to blank off the 50 mm diameter hole in the casing. Draw an undimensioned, half-sectional view through the assembled parts corresponding to the given view of the casing. Any dimensions not shown are to be settled by the student.

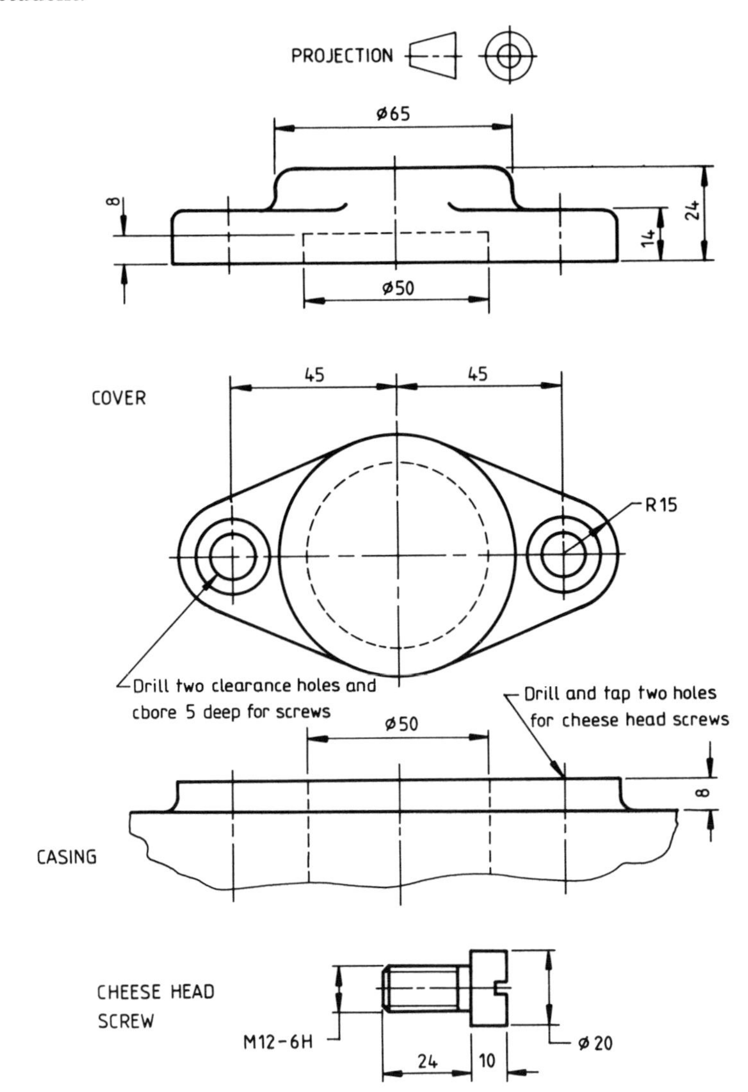

◀**Figure 12.7** Cover assembly

Learning Assignment 13

Dimensioning ISO screw threads

(a)

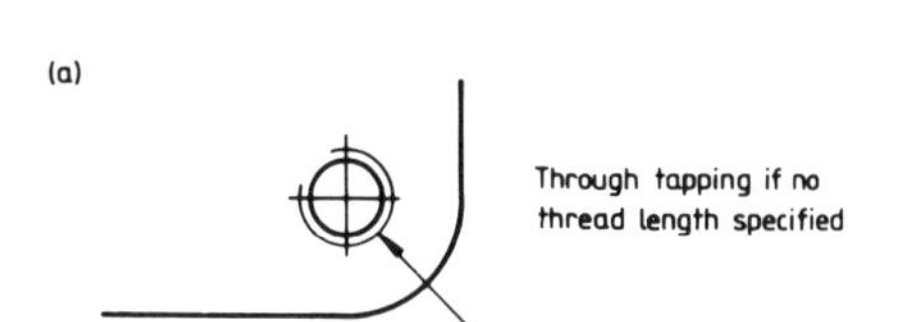

(b)

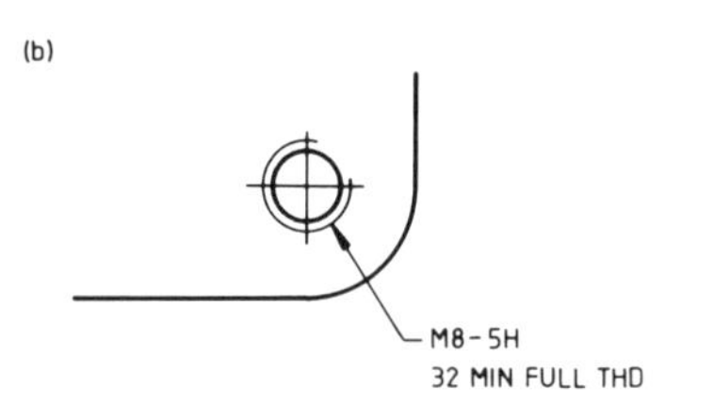

(c)

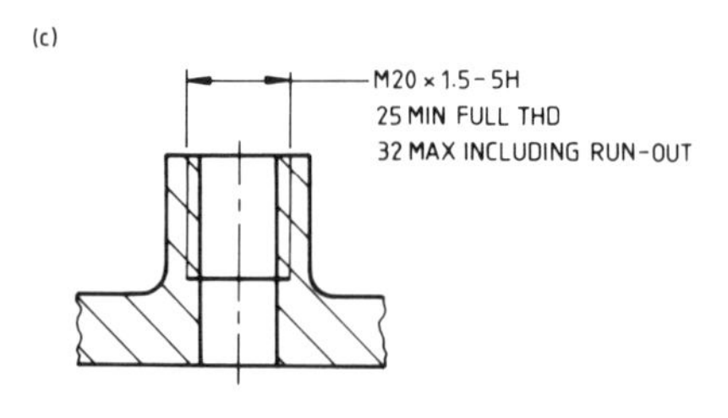

(d)

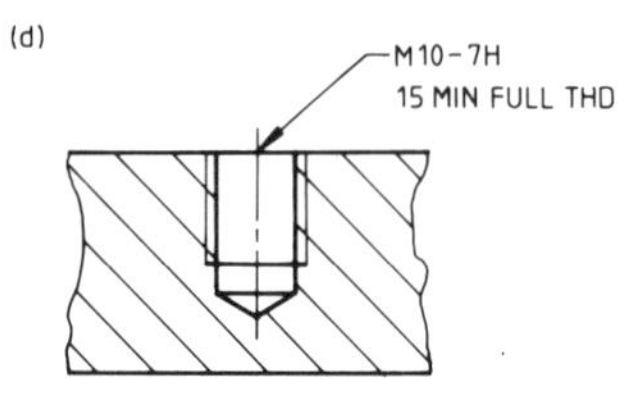

(e)

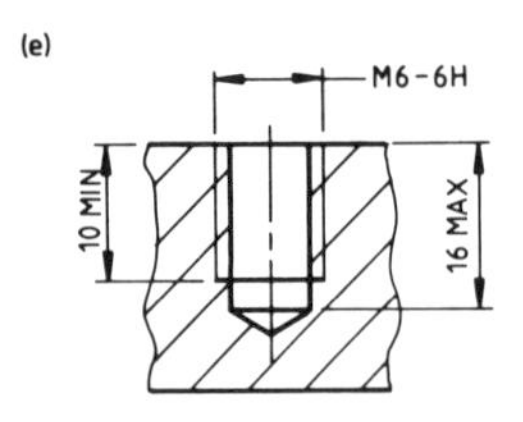

(f)

(g)

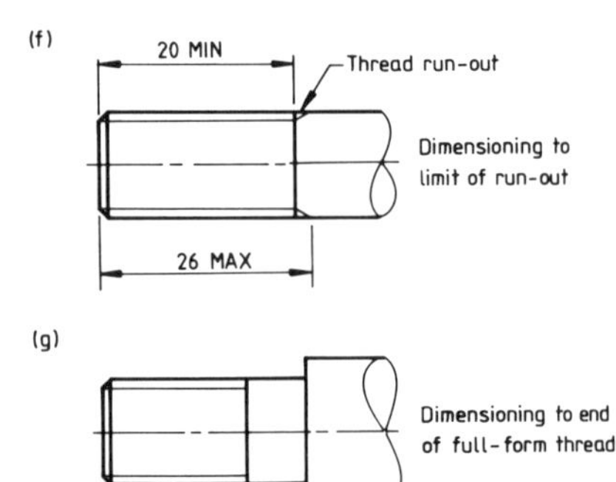

▲**Figure 13.1** Dimensioning ISO screw threads

Standard screw threads can usually be specified completely by a note. For special screw threads the major, effective and minor diameters may be given, as well as the pitch and the thread tolerance class (see below).

The ISO metric thread is the thread most commonly used, and is designated as in the following examples:

- For an internal thread: M8 × 1.25–6H.
- For an external thread: M20 × 2.5–6g.

M is the symbol for the ISO thread system; 8 and 20 are the major diameters in millimetres; 1.25 and 2.5 are the pitches of the threads in millimetres; 6H and 6g are the symbols for the thread tolerance classes. The tolerance class symbol is the only difference between the designations for internal and external threads. Following the usual convention for designating limits and fits, given in Assignment 11, capital letters designate internal threads and lower-case letters external threads, but whereas the letter in the symbol normally precedes the number, for threads the order is reversed.

The pitch of a thread is the distance from a point on one thread form to the corresponding point on the next thread form. It is illustrated in Figure 12.1 on page 63, together with other screw thread terms.

Additional information

Full details of the ISO thread are given in BS 3643. Pipe threads, which may be parallel or tapered, are used in certain branches of engineering. BS 21 covers the parallel type and BS 2779 the tapered.

ISO thread pitch series

The ISO thread system has two pitch series: coarse and fine. By convention, the pitch may be omitted from the designation of a thread from the coarse-pitch series. So the designation of a coarse-pitch thread M6 × 1–6H may be abbreviated to M6–6H.

Additional information

Coarse-pitch threads are used almost exclusively. Fine-pitch threads are used mainly in special applications such as on fine-adjustment machine tools and where threads have to be formed on thin-walled components.

Fits for ISO threads

Three classes of fit, medium, close and free, are provided by the internal and external thread tolerances. The medium fit (6H/6g) is used for most general engineering applications. When high accuracy of thread form and pitch is needed the close fit (5H/4h) is used. The free fit (7H/8g) is suitable when quick and easy assembly is necessary, even if the threads are dirty or slightly damaged.

Thread designations applied to drawings

Figure 13.1 gives examples of how thread designations should appear on drawings. Figure 13.1 (a) and (b) show how the designation is applied to the circular view of an internal thread. Methods for dimensioning sectional views of internal threads are given in Figure 13.1(c), (d) and (e). Figure 13.1(f) and (g) illustrate how the full-form thread length and the run-out length for an external thread may be controlled where necessary.

If the thread is left-handed then the abbreviation LH follows the designation. No indication of the hand is needed for a right-handed thread.

The thread run-out need not be shown on any thread unless it is essential for the part to function correctly. An example is the run-out threads on a stud. These are screwed into the tapping until they jam in the first thread in the hole, so securing the stud. The conventional representation for a run-out is given in Figure 13.1(f). It is drawn using a thin line at 30° to the major diameter.

A POINT TO NOTE

If the tapping drill for an internal thread is not to pass right through the part, make sure that the depth of the drilling is specified.

Additional information

The dimensions for undercuts and run-outs for screw threads are given in BS 1936.

TASKS

Task 13.1

Draw the given left-hand view of the threaded bracket in Figure 13.2, replace the right-hand view with section A–A and add section B–B.

(a) At P show three M8 medium-fit, coarse-pitch, through tappings.
(b) At Q show an M20 medium-fit, coarse-pitch thread 25 mm deep with a tapping drill depth of 36 mm. A 32 mm diameter spotface is required at the start of the thread.
(c) At R show an M24 close-fit, fine-pitch, through thread with a 35 mm diameter counterbore 12 mm deep at each end.

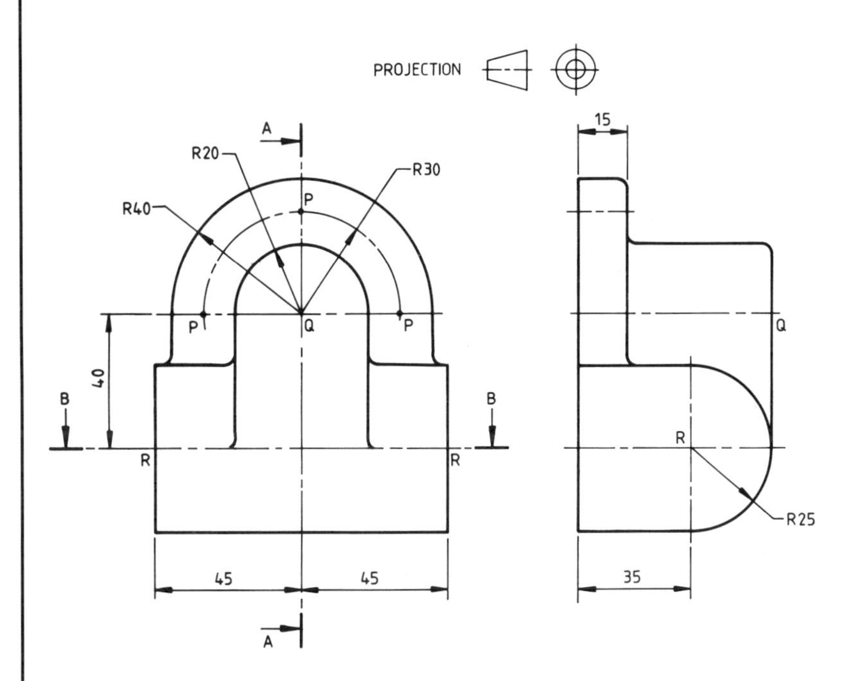

◀Figure 13.2 Threaded bracket

Do not show any dimensions except to specify the threads, spotface and counterbores completely.

Fillets and rounds are to be R3.

Task 13.2

Figure 13.3 shows the parts for a special drive assembly. The part shown at (a) has M10 tappings through the flange at A, B, C and D. Threaded pins shown at (c) are screwed into these tappings, and are locked by tightening the thread run-out into the first thread in the tapping.

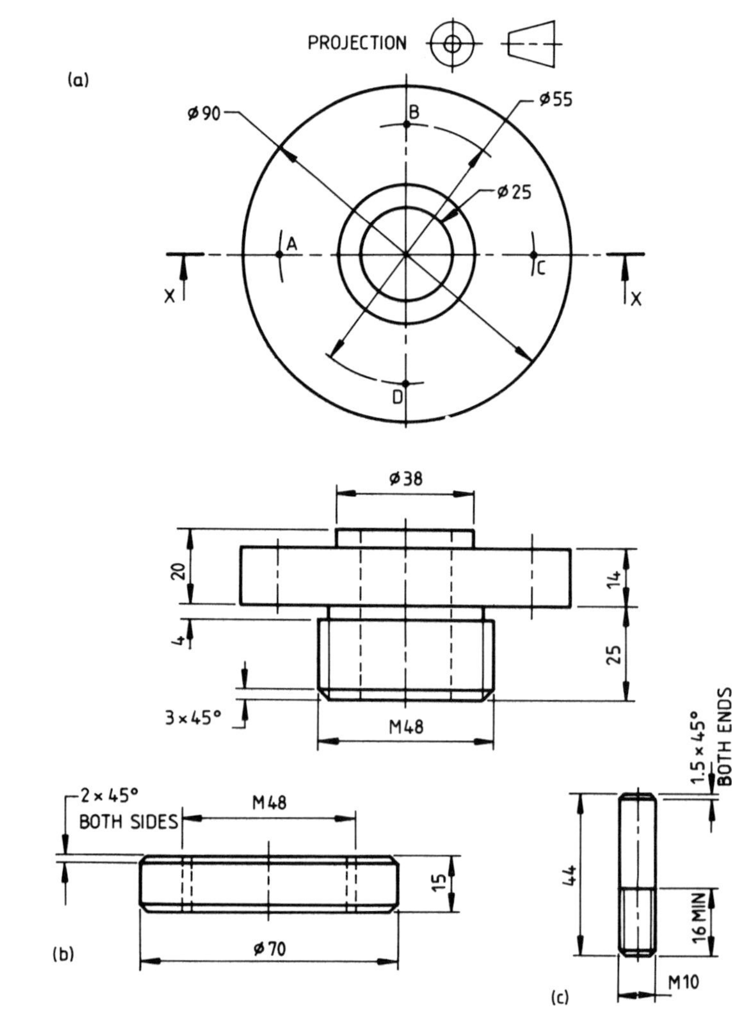

►**Figure 13.3** Special drive assembly

Draw the top view of part (a), with the tappings, but showing only the pin at A in position. Add section X–X with the ring nut, part (b), loosely assembled. Dimensions are not required.

The threaded pins are to be a close fit in the tappings and the ring nut is to be a free fit on part (a). Write down the designations for each thread, if they are all from the coarse-pitch series.

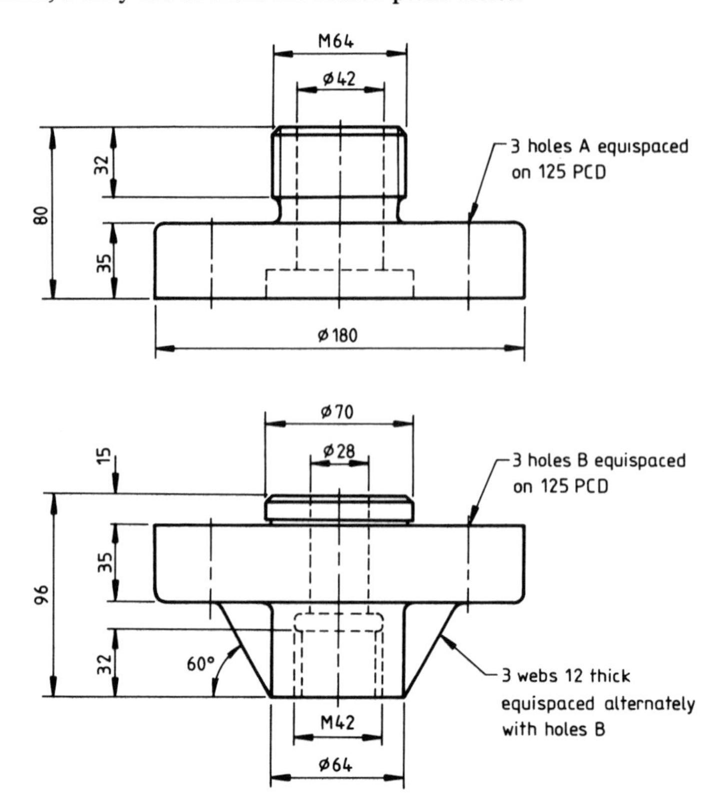

►**Figure 13.4** Coupling assembly

Task 13.3

The parts of the coupling assembly shown in Figure 13.4 are to be secured together by three M20 hexagon bolts and nuts. Holes A and B are clearance holes for the bolts, and the bolt heads are to be completely housed in counterbores machined in holes A.

Draw suitable views of both parts. Any omitted dimensions are to be settled by the student. Show the following dimensions only, toleranced as necessary:

(a) Those for the counterbores and clearance holes. For the counterbore diameter use a tool manufacturer's catalogue to find the outside diameter of an M20 hexagon socket and then find the diameter of the nearest larger counterboring cutter. The depth of the counterbore should be a little greater than the thickness of an M20 bolt head. Find this from a fastener manufacturer's catalogue or from BS 3692. The diameter of the clearance holes can be found from BS 4186, which provides three classes of fit. Select the most appropriate. Alternatively, your company may have its own standard.

(b) Those for the locating spigot and recess, having decided on a suitable fit.

(c) A suitable diameter for a spotface under the nuts. Use a fastener catalogue or BS 3692 to find the across-corners dimension of an M20 nut and use this to select the next larger spotface cutter.

(d) Those for the M64 and M42 threads and their associated undercuts. The threads are both from the coarse series and are to be medium fits with the mating parts. Use BS 1936 or your company's standard to settle the undercut dimensions.

Learning Assignment 14

Conventional representation of common features

It is often forgotten that the total cost of making a part includes the cost of making the drawing for it. So any method of keeping the cost of the drawing to a minimum should be used. One way of doing this is to employ the various conventional representations which are available. These reduce the time needed to complete the drawing and some also save space. In this event it might be possible to draw a view to a larger scale, so making the drawing easier to read. On the other hand, a smaller size of drawing sheet could be used. The smaller the sheet the cheaper it is to buy, store, transmit and reproduce.

A number of conventional representations are recommended in BS 308, as well as that for screw threads, which is covered in Assignment 12. Those which follow are the most commonly used.

Interrupted views

Long parts whose cross-section is the same shape throughout their length can have a piece removed from the centre of the view which shows the length. The ends of the view, sufficient to define the part, are then drawn close together as partial views, as in Figure 14.1, thus saving space on the drawing.

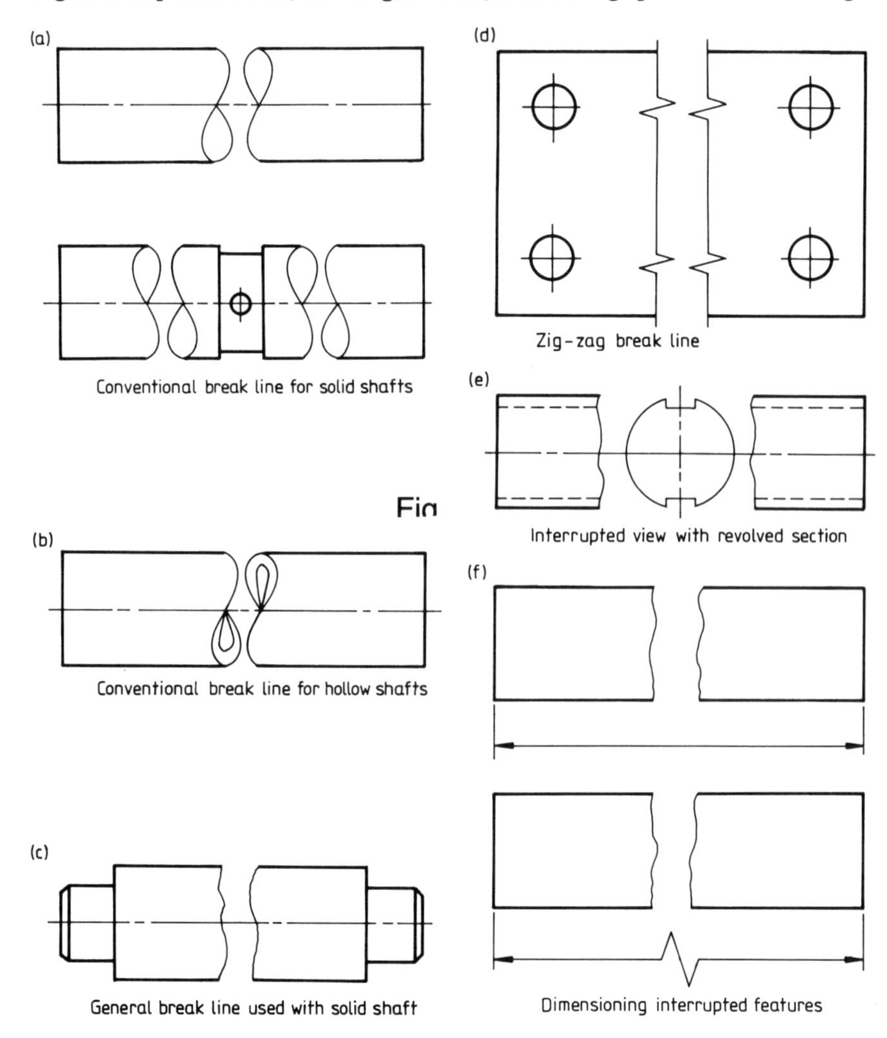

▶Figure 14.1 Interrupted views

Figure 14.1(a) and (b) illustrate the conventional break lines for solid and hollow shafts on interrupted views. Alternatively, these may be drawn using the general break line in Figure 14.1(c). Figure 14.1(d) shows a break line which can be easily drawn with a CAD machine. Note that break lines are thin.

Sometimes a revolved section is drawn between the parts of an interrupted view as in Figure 14.1(e).

Dimension lines for interrupted features should not be broken. Two ways of showing them are given in Figure 14.1(f).

Interrupted views, using the break lines in Figure 14.1(c) and (d), can also be used on sections.

Rolling bearings

This conventional representation, illustrated in Figure 14.2, can be used on assembly drawings for all types of ball and roller bearings. If the internal details of these bearings had to be shown when they appeared on sectional views, then the races, cages and balls or rollers would all have to be drawn. This would be very time-consuming and unnecessary, because these parts are not dimensioned on assembly drawings and the bearings are identified by a note.

The central cross is drawn with thin lines which intersect at the centre of the cross-section of the bearing. The length of the lines of the cross is half the width of the cross-section.

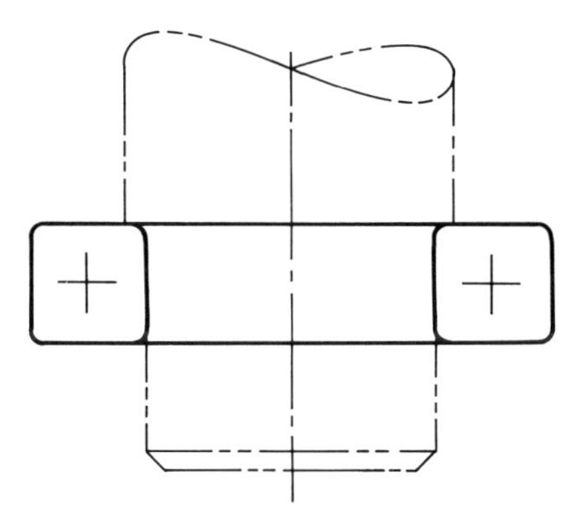

▲**Figure 14.2** Conventional representation of rolling bearings

Flat surfaces on cylindrical parts

Flat surfaces such as squares and local flats on shafts and similar parts may be indicated by thin diagonal lines, as shown in Figure 14.3.

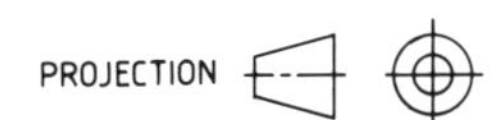

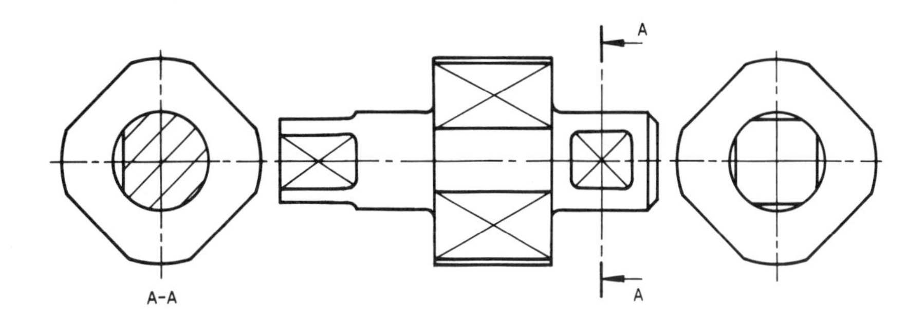

◀**Figure 14.3** Conventional representation of flat surfaces on cylindrical parts

> *POINTS TO NOTE*
> - The lines representing the knurling are thin and the spacing between them is quite wide to avoid them closing up on reduced-size prints.
> - For diamond knurling the lines are at 30° to the outside diameter of the cylinder to which the knurling is to be applied.
> - The conventional representation is applied to a small area only of the knurled surface.

Knurling

Knurling roughens the surface of the heads of adjusting screws, handles and similar parts to provide a better grip. There are two types: straight knurling and diamond knurling, and Figure 14.4 gives the conventional representations for both.

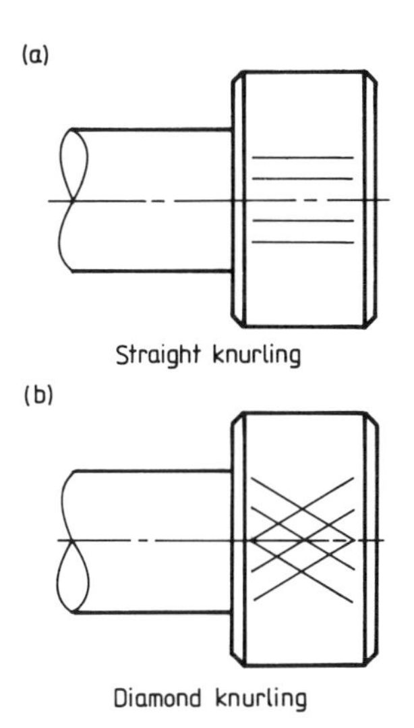

▲**Figure 14.4** Conventional representation of knurling

TASKS

Task 14.1

Draw only the given interrupted view of the jig pin in Figure 14.5, adding:

- the break lines and revolved section;
- the visible flat;
- the diamond knurling.

The appropriate conventional representations are to be used. Do not show any dimensions.

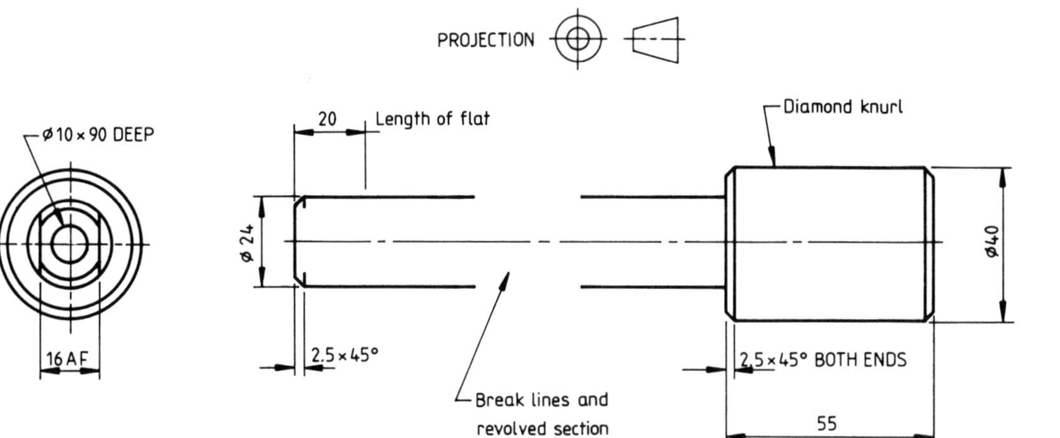

▲**Figure 14.5** Jig pin

TASKS

Task 14.2

Figure 14.6 is the assembly drawing for a reciprocating mechanism, together with a view showing the arrangement of the shaft, which is driven through the V-belt pulley, and the link and rod. Rotation of the shaft causes the top of the rod (point D) to rise and fall.

Fully dimensioned drawings are required of the casing, shaft, link, rod, pivot pin and bushes. The end-float of the shaft is to be between 0.15 mm and 0.65 mm and the tolerance on the rise and fall of the end of the rod is to be 0.4 mm. Suitable fits are to be specified where necessary. Any omitted dimensions are to be decided by the student.

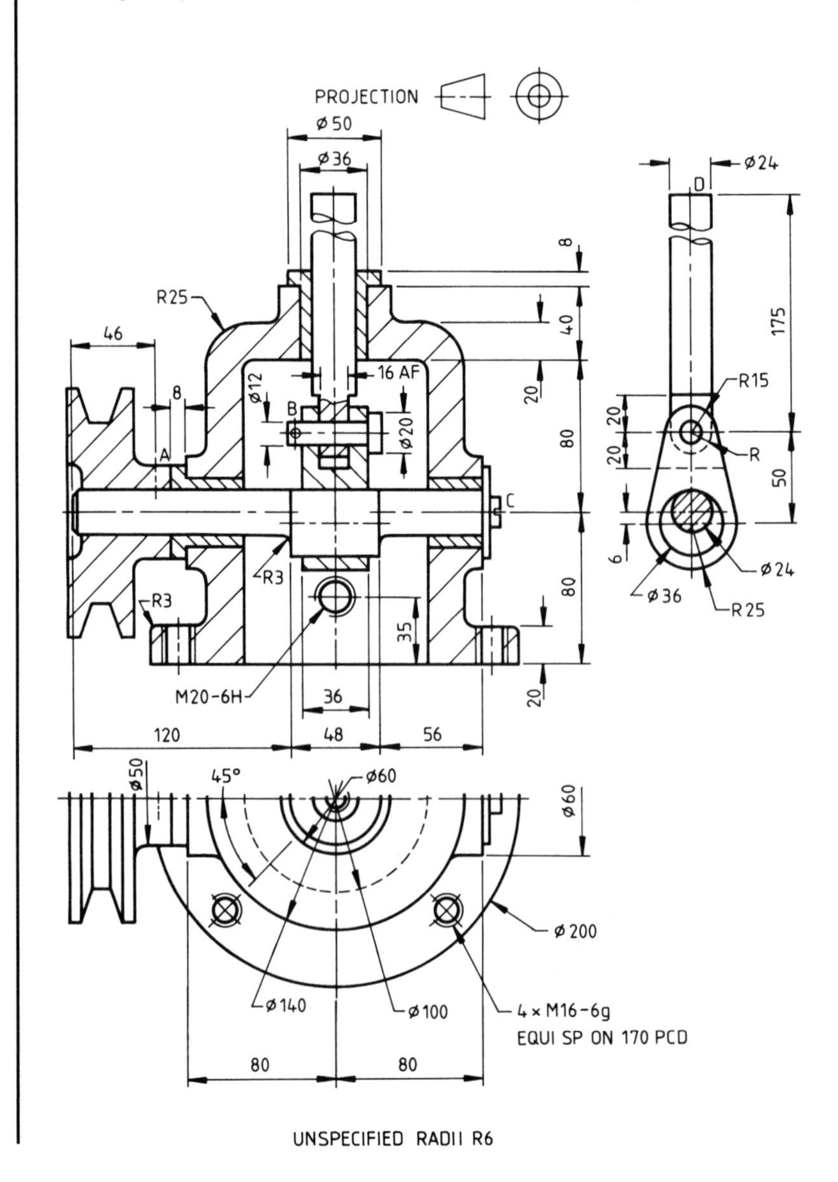

►**Figure 14.6** Reciprocating mechanism

The pulley is secured to the shaft by an M8 hexagon socket set screw at A. The screw has a dog point which engages with a shallow hole drilled in the shaft.

At B a split pin retains the pivot pin in the link and rod. An M8 cheese head screw, 18 mm long, in an axial tapping in the shaft at C, retains the washer against the end of the shaft.

At D there is an M6 medium fit, coarse-pitch axial tapping with a full thread depth of 10 mm. On the outside of the rod at the end is a pair of flats 20 mm apart and 25 mm long.

For information on the set screw and split pin refer to British Standards, fastener manufacturers' catalogues, or your company's standards.

Learning Assignment 15

Machining and surface texture indication

All manufactured surfaces, however they are produced, have imperfections. These take the form of a series of peaks and valleys which vary in height and spacing. They result in a texture for the surface which in feel and appearance is generally characteristic of the process which produced it.

Controlling the texture of a surface can lead to longer life for the component, fatigue resistance, improved efficiency and functional interchangeability. Control is exercised by specifying a surface texture of known type and roughness value which experience shows is the most suitable to secure these benefits at the lowest cost. Roughness values are expressed in micrometres (μm) and given the symbol R_a. They are the average of the heights and depths of the peaks and valleys from a reference line through the profile of the surface. Preferred values are 50, 25, 12.5, 6.3, 3.2, 1.6, 0.8, 0.4, 0.2, 0.1, 0.05, 0.025 and 0.0125 μm.

A POINT TO NOTE

To keep the cost of a part to the minimum, the coarsest roughness value which is functionally acceptable should be specified.

Typical applications of preferred surface roughness values

50 μm and 25 μm

Surfaces whose appearance is unimportant, which are not members of a fit and not subject to stress concentration, vibration or fatigue.

12.5 μm and 6.3 μm

Non-fitting surfaces with wide tolerances not subject to stress concentration, vibration or fatigue. Used on surfaces of bolted and riveted joints on structural steelwork, shaft diameters and ends, undercuts and faces of pulleys and couplings. Not suitable for accurate location faces.

3.2 μm

Gives the surface a reasonable appearance. Maximum surface roughness for non-fitting surfaces when low stress concentration and fatigue are present. Used for general machining and location surfaces with large tolerances, drilled holes and joint faces for use with soft gaskets. Does not provide an air- or oil-tight seal with metal-to-metal joints.

1.6 μm

Non-fitting surfaces subject to low to medium stress concentration or fatigue. Used for location surfaces with closer tolerances, slideways and beds, and slow-running plain bearings for light duty.

0.8 μm

In general use for press fits and plain bearings. Also used for surfaces subject to stress concentration, mating surfaces of cylinder heads and blocks, rolling bearing seatings in housings and on shafts, grooves for static O-rings and brake drums.

0.4 µm

Used for rolling bearing rings and seatings for precision fits, precision journal bearings, gear teeth, and mating faces for clutches. Also used for parts which slide or rotate in contact with packings and oil seals. Air- and oil-tight under pressure with metal-to-metal joints.

0.2 µm

Used for highly stressed journal bearings in internal-combustion engines, rotating or sliding hydraulic surfaces, surfaces in sliding contact with O-rings, piston valves, ram cylinders and rolling bearing races.

0.1 µm

Used for high quality rolling bearings, precision hydraulic parts, face seals and shop gauges.

0.05 µm

Used for reference gauges and comparator anvils.

0.025 µm

Used for slip gauges and master reference gauges.

Machining and surface texture symbols

The more common machining and surface texture symbols are shown in Figure 15.1.

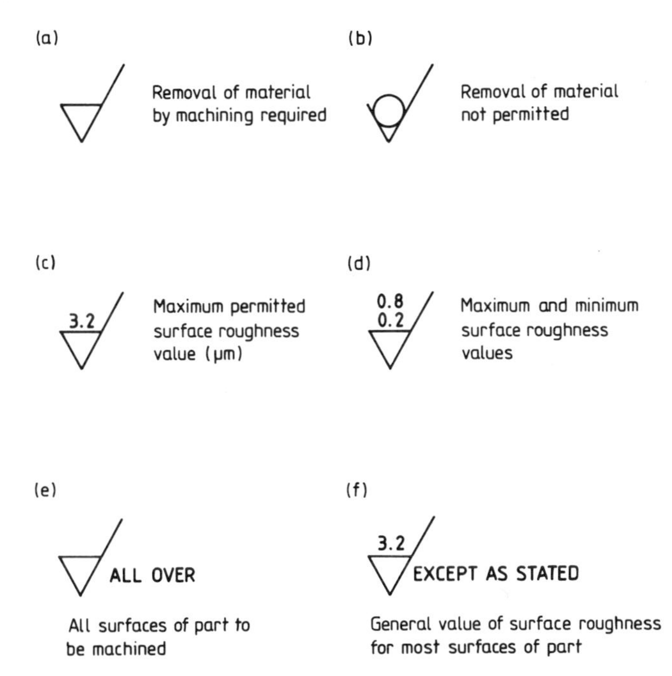

◀**Figure 15.1** Indications of machining and surface texture

When a surface is to be machined the symbol in Figure 15.1(a) is used. Figure 15.1(b) requires that no material is removed from the surface by machining. Surface texture values can be added to the machining symbol if necessary, as shown in Figure 15.1(c) and (d). If maximum and minimum values are shown, as in Figure 15.1(d), then the maximum value is placed above the minimum.

When all the surfaces of the component are to be machined then the symbol in Figure 15.1(e) is added to the drawing. It is usually placed near the title block. A surface texture value may be applied to the symbol if required.

If the same surface texture value applies to several surfaces of the component the machining symbol and value may be stated in a note. Figure 15.1(f) is an example. The note usually appears near the title block.

Application of symbols and values to drawings

A POINT TO NOTE

The line thickness of the machining symbols should be the same as that used for the lettering on the drawing.

Generally machining symbols should be shown once only on a surface, preferably on the view which shows the dimensions which define the surface. Acceptable positions for the symbols are given in Figure 15.2.

Surface texture values added to machining symbols should be placed so that they can be read from the bottom or the right-hand side of the drawing, as with the ordinary dimensions on the drawing. See Figure 15.2.

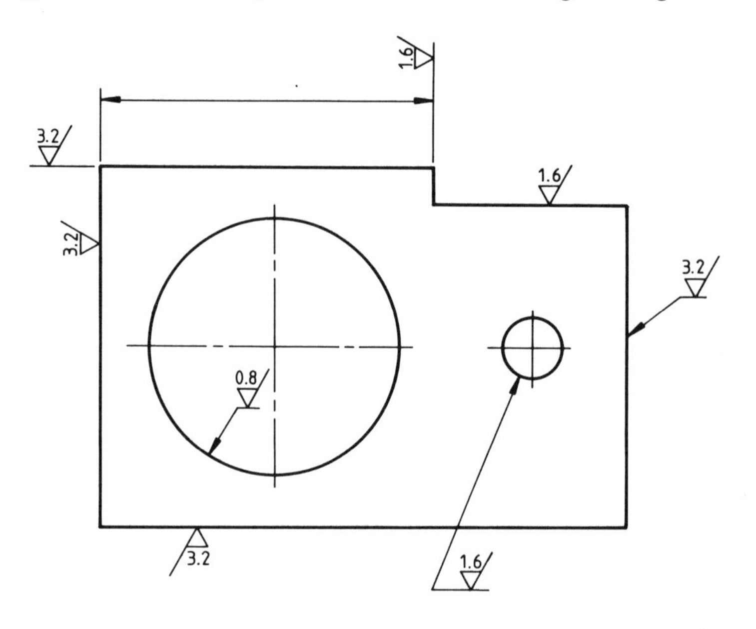

▶Figure 15.2 Application of machining symbols and surface texture values

Additional information

Average surface roughness values for some common production processes are:

Flame cutting	25 to 12.5 µm
Sawing	25 to 3.2 µm
Planing, shaping	25 to 0.8 µm
Drilling	6.3 to 1.6 µm
Milling	6.3 to 0.8 µm
Broaching, reaming	3.2 to 0.8 µm
Boring, turning	6.3 to 0.4 µm
Grinding	1.6 to 0.1 µm
Lapping	0.4 to 0.05 µm
Sand casting	25 to 12.5 µm
Hot rolling	25 to 12.5 µm
Forging	12.5 to 3.2 µm
Permanent mould casting	3.2 to 1.6 µm
Investment casting	3.2 to 1.6 µm
Extruding, drawing, cold rolling	3.2 to 0.8 µm
Die casting	1.6 to 0.8 µm

TASKS

Task 15.1

The cover plate shown in Figure 15.3 is 120 mm long by 80 mm wide by 15 mm thick. The large hole is 40H7 diameter and centrally placed in the plate. The four small holes are 12H9 diameter with their axes 12 mm from the plate edges. The datum feature for these axes and the plate edges is the axis of the large hole. All the size dimensions for the

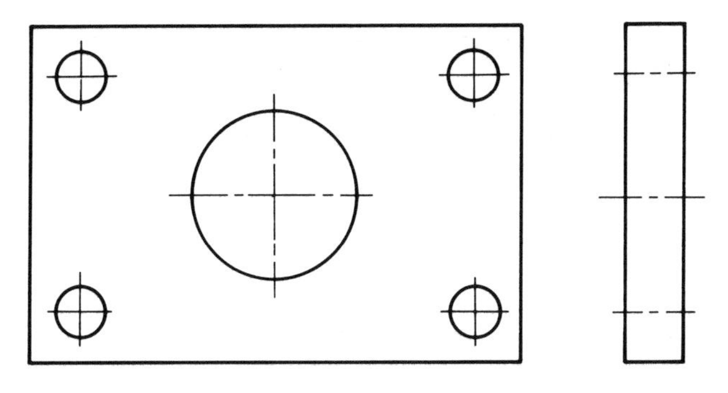

◀**Figure 15.3** Cover plate

plate carry limits of tolerance of 0 / –0.1. The location dimensions for the four holes have limits of tolerance of +0.08 / +0.1.

With the exception of the surfaces of the holes, all the plate surfaces are to have a maximum surface roughness value of 3.2 μm. The surface roughness of the large hole is to lie between 0.4 μm and 0.1 μm, and the value for the small holes is to be 1.6 μm.

Draw the given views of the plate and fully dimension them.

Task 15.2

Figure 15.4 shows the assembled parts for a roller guide. Make fully dimensioned and toleranced drawings of the mounting bracket, bush, shaft and guide, selecting suitable fits for the mating parts. The end-float of the guide and bush sub-assembly is to be between 0.3 mm and 0.45 mm. Show machining indications and add suitable surface roughness values where appropriate. Any omitted dimensions are to be settled by the student.

Consult manufacturers' catalogues, British Standards or your company's standards to settle the dimensions for the spotfaced holes in the mounting bracket and the tapping for the grease nipple, and to find the thicknesses of the M24 nut and plain washer.

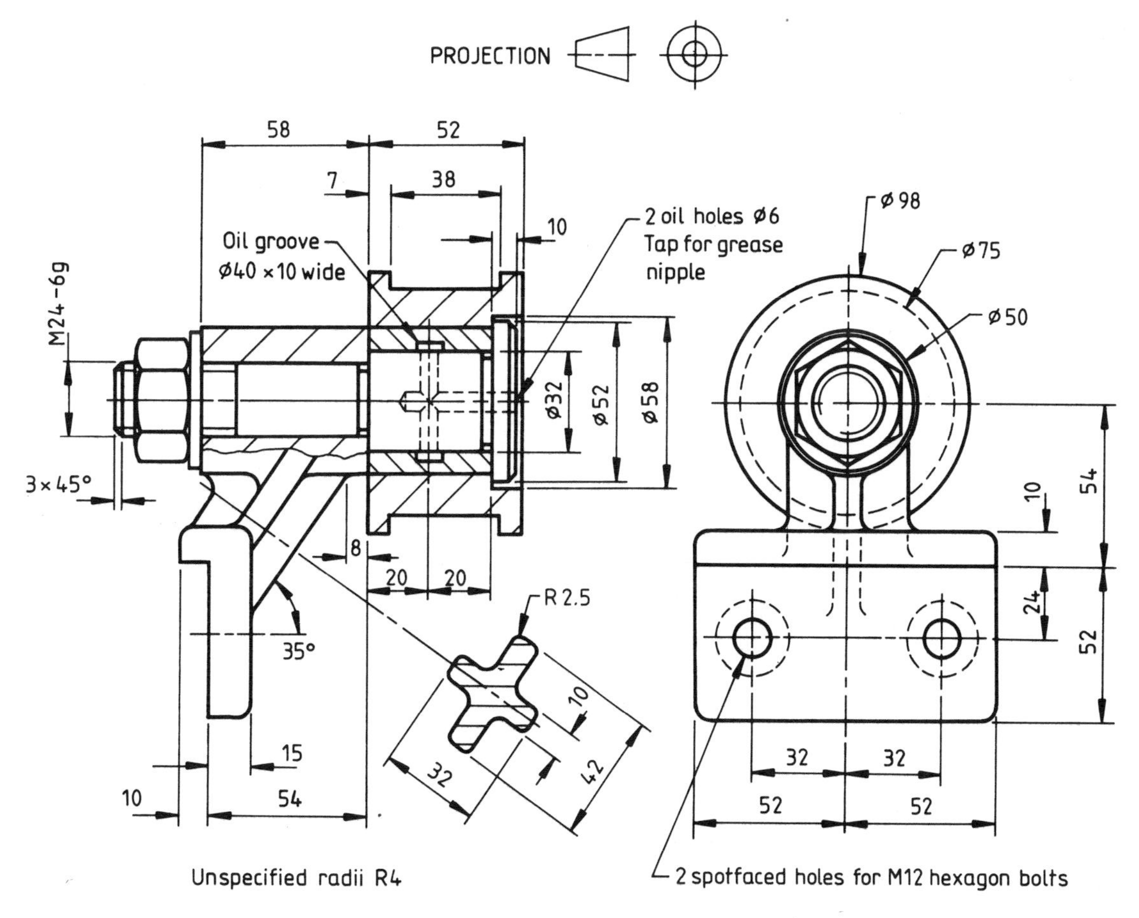

▲**Figure 15.4** Roller guide assembly

Index